Electricity Wayleaves, Easements and Consents
Litigation Practice and Procedure

Charles Hamer and Gary O'Brien

2007

Routledge
Taylor & Francis Group

LONDON AND NEW YORK

First published 2007 by Estates Gazette

Published 2014 by Routledge
2 Park Square, Milton Park, Abingdon, Oxon OX14 4RN
711 Third Avenue, New York, NY 10017, USA

Routledge is an imprint of the Taylor & Francis Group, an informa business

© Charles Hamer and Gary O'Brien 2007

All rights reserved. No part of this book may be reprinted or reproduced or utilised in any form or by any electronic, mechanical, or other means, now known or hereafter invented, including photocopying and recording, or in any information storage or retrieval system, without permission in writing from the publishers.

Notices
Knowledge and best practice in this field are constantly changing. As new research and experience broaden our understanding, changes in research methods, professional practices, or medical treatment may become necessary.

To the fullest extent of the law, neither the Publisher nor the authors, contributors, or editors, assume any liability for any injury and/or damage to persons or property as a matter of products liability, negligence or otherwise, or from any use or operation of any methods, products, instructions, or ideas contained in the material herein.

ISBN 978-0-7282-0505-5 (pbk)

Cover design by Ted Masters Typeset in Palatino 10/12 by Amy Boyle, Rochester

Contents

Acknowledgements viii

Preface ix

Introduction
1. Electricity: Power Without Responsibility? 3

Part 1 — The Electricity Industry
2. Electricity Industry Structure 9
3. Electricity Infrastructure 15
4. Energy Legislation Since 1989 25

Part 2 — Private Rights
5. Non-Statutory Permissions Granted by
   Private Agreement 33
6. Non-Statutory Wayleaves: Voluntary Wayleaves 51
7. Non-Statutory Wayleaves: Implied Wayleaves 59

Part 3 — Public Control
8. The Statutory Menu Relating to Wayleaves,
   Consents and Easements. 65
9. Applications to the Secretary of State for
   Statutory Rights 73
10. Necessary Wayleave Hearings. 87
11. Compulsory Purchase Powers. 105
12. Planning Permission and Ministerial Consent 113
13. Public Inquiries. 121

Part 4 — Financial Aspects
14. Wayleave Payments 133
15. Easements 143
16. Applications to the Lands Tribunal 157

Part 5 — Enforcement of Rights
17. Relevant Remedies — An Overview..................... 209
18. Failure to Comply with Financial Obligations 217
19. Refusal to Allow Exercise of Wayleave Rights 223
20. Compliance with Notice to Remove..................... 227
21. Wrongful Use of Land by Licence-Holder................ 231

Conclusion
22. Utility Wayleaves — The Need for Reform............... 237

Appendices
1. Map of Area Electricity Boards 1947–1990................ 249
2. Map of Distribution Network Operators 2006 251
3. Table of Area Electricity Boards, Regional Companies and Distribution Network Operators..................... 253
4. Illustration of the Generation, Transmission and Distribution of Electricity 255
5. Typical Suspension/High Voltage Towers................. 257
6. Typical Double Circuit Overhead Line.................... 259
7. Typical Single Circuit Overhead Line..................... 261
8. Glossary of Terms Relating to Electricity Apparatus 263
9. Clearance to Trees 269
10. Clearance to Objects................................... 271
11A. Flowchart of the Necessary Wayleave Process 273
11B. Contents of an Application for a Necessary Wayleave: General Requirements................................. 277
11C. Contents of an Application for a Necessary Wayleave: Additional Information in Respect of an Existing Line 279
11D. Contents of an Application for a Necessary Wayleave: Additional Information in Respect of a New Line 281
12. Form B (Type II) 283
13. Form B (Type III) 289
14. Modifications to Overhead Line on Existing Route 295
15. Permanent Overhead Line Diversions 297
16. Temporary Overhead Line Diversions 299
17. Permitted Development by Licence Holders.............. 301

18.	Wayleave Payments	305
19.	Electricity Wayleave Payments to Landowners	307
20.	Electricity Industry Payments to Occupiers Explanatory Notes	309
21.	Tower Size Allocations for Wayleave Payments	311
22.	Lands Tribunal Form R	313
23.	Notes on Completing Form R	321
24.	Notice Periods and Dates of Service	327
25.	Relevant Acts of Parliament	335
26.	Relevant Statutory Instruments	337
27.	Comparison of Utility Legislation in Respect of Bases for Compensation	339
28.	Useful Addresses	341
Index		343

Acknowledgements

Our thanks go to Tina Wing and Amy Dunn for their patient typing of the text through myriad permutations. Our thanks go also to Nazia Hussain, Liz Turner, Laurence Gray, Robert Phillips, Paul Moorcroft, Lisa Barge, David Feist, Nick McKnight, Chris Jones, Hannah Clayton, Collette Davies, Natalie Burns, Helen Worsnop, Sally Milliner and Stacey Boston all of Eversheds LLP who took on the task of proof reading the manuscript. Within Hamer Associates, particularly for the carrying out of additional research, we are indebted to Vicki Thorpe and Nick Marshall for their considerable assistance.

Finally, our thanks must go to our families who have tolerated long absences from the home while we have been working on this project. We may not be applauded by lengthy queues of eager purchasers outside every high street retailer but the experience of combining the outlook and collective experience of surveyor and solicitor has been mutually beneficial and rewarding.

Preface

In our introductory chapter we draw attention to the changing attitudes to be found among landowners and electricity companies in equal measure. The newer approaches adopted by each interest group have prompted an increase in the number of disputes relating to electricity wayleaves. It is for that reason that we concluded that the time was ripe for our book to appear.

The policy questions which form the backdrop to our book are succinctly summarised in Chapter 1 as follows:

> The drive fully to develop land at a time when increased housing density is a vital component of government policy continues unabated. The need for larger and more efficient commercial buildings requires that available and consented land is properly developed. This trend for increased development and the changes that have taken place in the UK electricity industry make conflict between full land use and the presence of electricity infrastructure increasing likely.

These policy questions achieved even greater prominence with the publication on 23 May 2007 of the Government's White Paper: "Meeting the Energy Challenge". The drive towards a low carbon economy depends on the generation of energy from renewable resources. It also depends on electric lines for the transmission and distribution of energy so produced. Producers of such energy must still connect to the national grid if the link between generation and supply to the customer is not to be broken. Given the wide range of issues considered by the White Paper, we would have been content to leave matters there had it not been for the publication on 27 April 2007 of a crucial report by SAGE. SAGE is the acronym for the stakeholder advisory group on extremely low frequency electric and magnetic fields. The report is entitled "'Precautionary approaches to ELF EMFs"'.[1] Unfortunately the SAGE report entered the public domain too late for us to make any substantive changes to Chapter 1. However some commentary is undoubtedly required. Following the example of Hollywood blockbuster franchises we have penned this prequel as the vehicle for our observations on the SAGE report which first we must put into context by expanding on the quotation above.

Few of us will realise that every time we turn on a switch we are playing a small but nevertheless important role in a much larger policy debate. You might think that this refers to the debate about sources of fuel and renewable energy. In fact we are alluding to the debate about housing need.

The 6th Report of the Parliamentary Select Committee on Transport, Local Government and the Regions [20 March 2002] addressed the problem of empty homes, describing this as an "'issue of national importance"'. The executive summary of the report has this comment

> If the fortunes of inner urban areas are to be turned around, they need to become attractive places to live. In northern and midlands cities with growing economies there is the potential for large scale, conurbation-wide interventions which include the reconfiguration of large areas to make them more attractive, not only to existing residents but to new residents. To be successful, such conurbation-wide interventions will need new powers and new ways of working with significant additional funding.

The regeneration of existing housing stock and the stimulation of demand for it given that such stock tends to be in our traditional industrial heartlands would clearly have to grapple with the impact of electric lines. Proposals for new housing likewise will have to take electric lines into account insofar as new housing is going to be constructed over former agricultural land. As we explain in Chapter 1 it is over such land that much electrical apparatus has been erected.

The government's housing strategy is not confined to the regeneration and re-use of empty dwellings. The South East Regional Strategy 2006–09 which went out to consultation between January and March 2005 sought to guide the government's allocation of housing funding for the years 2006–07 and 2007–08. Nationally, an extra £1.3 bn

---

1   ELF EMF refers to extremely low frequency electric and magnetic fields as covered in the terms of reference in the SAGE report

was said to be available to fund attempts to alleviate the housing shortage over that period. Also in January 2005 the office of the Deputy Prime Minister published a five year plan in a report entitled "'Sustainable Communities: Homes for all"'. The report is described as the "'next phase"' in delivering the "'Sustainable Communities Plan"' which included the following aims:

- Working in partnership to deliver 1.1 million new homes in the wider South East by 2016 and investing in new jobs and infrastructure
- Enhancing the environment -with new powers to limit low density development and protect the green belt-and a new Code for Sustainable Buildings
- Extending our £1.2 billion market renewal work in the North and Midlands to new areas and piloting new approaches to encourage mixed communities in run down estates and communities (sic)

Against this policy background SAGE's report has been released. The SAGE process was initiated by National Grid but is now under the lead of the Department of Health. It is funded by the Department of Health, National Grid and the Energy Networks Association and the charity Children with Leukaemia. The full report is available via the web-site of the consultants who have facilitated it: RK Partnership Ltd the address of which is *www.rkpartnership.co.uk*.

The publication of this report provides a timely reminder of the need for well informed advice on electricity wayleaves. The SAGE report considers the effects of overhead power lines in proximity to residential development and makes numerous references to wayleaves and associated compensation issues without being able to address them in any detail.

The report concludes with options for government to consider. We wish to quote the following extract from the conclusion which will be of particular interest to readers of this book and explains why we thought it desirable to write this preface:

> We have identified two precautionary measures (better information for members of the public and optimal phasing of 132 kV overhead lines not already thus phased) that we recommend. These options, however, will not have a dramatic effect on exposures [to extremely low frequency electric and magnetic fields]. We have therefore identified the best-available option for obtaining significant exposure reduction (in fact, it avoids new future exposures that would otherwise occur). This is a restriction on new homes and schools close to existing [electric] lines, and on new lines close to

existing homes and schools (the "corridors for new build" option, which can in fact be seen as a suite of options used to achieve a single objective). The main costs of this option arise from the effects on land and property values. We urge government to make a clear decision on whether to implement this option or not.

Whether the government does so act remains to be seen. It is not the purpose of this book to analyse or contribute to the development of policy. Still less is it the intention of the authors to offer a view on the merits of these plans. Our final chapter does, it is true, make proposals for the reform of the law on this subject so as to facilitate more economic and effective disposal of disputes. Nevertheless, this is a practitioner's guide. What we can suggest is that until the government does so act, it is all the more likely that both landowners and electricity companies will seek to use their private law remedies to negotiate their way through the complex issues which arise.

As if all these changes were not enough the electricity industry itself (as the White Paper makes abundantly clear) is facing significant challenges. The initial concept of a national grid in the 1930s was further developed with the growth of demand for electricity in the 1960s requiring a supergrid to be established. The electricity network was designed largely around coal fired power stations in the north and the midlands. The changing nature of power station fuel supplies, relying on gas from the north sea, and increasing demand in the south east of the country has come at a time when much of the electricity infrastructure is at or approaching 50 years old. Further connections and reinforcement of transmission and distribution lines and the costly replacement of time expired assets, much of which is likely to now be wrongly located, can only increase the extent to which landowners and electricity companies are going to have to come to terms against a background of conflicting rights and duties.

It is for that reason that our subtitle is "'Litigation Practice and Procedure'". With that final comment we invite the reader to proceed to Chapter 1 ...

Introduction

Electricity: Power Without Responsibility?

To adapt a famous opening sentence of Jane Austen: it is a truth universally acknowledged that a utility looking to expand its operations must be in want of a good site for its infrastructure. However, while everyone may generally agree that electricity, gas, water and telecoms are all essential preconditions to a healthy economy (to say nothing of Darcy-like standards of living) when it comes to the very specific question of where the infrastructure should actually be located, there is never one single universally accepted answer. It is the purpose of all legislation relating to the resourcing of utilities to balance the national need for public provision against your right to preserve your private property as you prefer it to be.

This book explains the various methods which the Electricity Act 1989 employs to enable the electricity industry to function having regard to the need for that balance. However, we must offer an early word of warning: those methods are very distinct from the ones adopted by law for water, gas and telecoms. It might be thought that Parliament would authorise a utility so vital as electricity to acquire property outright or to take substantial legal rights from a land-owner if it needed to place infrastructure on private land. Terms such as easement, statutory easement, compulsory purchase, and lease might even be thought to be relevant. Not so! Electricity infrastructure rests (legally speaking) on an arrangement unique to that industry called a wayleave.

Neither outright purchase nor the granting of leases or easements became the norm during the growth of the electricity industry.

Unfortunately, the terms easement and wayleave in particular are often interchangeable in common parlance. In fact, as will be explained they are very distinct concepts.

In recent years disputes over wayleaves have risen dramatically in intensity and in number, resulting in many delayed projects. In the last five years the number of applications made by licence-holders for compulsory rights has more than doubled: bearing testament to this.[1] It should be noted that these disputes will overwhelmingly concern existing established electric lines.

Proposals for new lines attract even greater opposition. The reaction of most landowners today on being requested to grant rights for a new overhead line, would almost certainly be an immediate refusal tempered by the assumption that compulsion would soon follow. This contrasts with the 1950s when rural electrification was greeted with enthusiasm by landowners in providing power to houses and farms. The rights for the lines would have received limited consideration and emphasis would have been on the placing of poles and pylons to minimise the disruption to agriculture.

This book has been written for the benefit of all those involved in the use and development of land in this specialised area: residential developers, industry corporates, property developers, land owners, private individuals, surveyors, architects, planning consultants and lawyers. The procedures laid down by the Act and other related statutes and statutory instruments are likely to have been encountered only a few times even by professionals; and therefore this book will set out the main principles and the procedures which are likely to be less well understood than in other specialist areas such as rights of light and party walls. Many people will have come across annual wayleave payments, either personally or when managing company's assets, but are likely to have had little regard to the basis behind them.

In this book we have opened out the term wayleaves to include the full range of relevant permissions including easements and statutory consents. The relevant legislation and the statutory rules are considered to enable the reader to approach this area of the law with greater confidence. We do not give specific legal advice. Each case must be treated on its own merits. Nevertheless the information will enable a professional to give a more considered opinion when advising a

---

1   Source DTI (238 IN 01/2 and 554 IN 05/6).

client and the client to ask more informed questions. Where we think that we can usefully suggest best practice, we do so.

A need for well informed and experienced advice to any landowner and electricity company has never been more critical. Large areas of land are restricted by electric lines. The drive fully to develop land at a time when increased housing density is a vital component of government policy continues unabated. The need for larger and more efficient commercial buildings requires that available and consented land is properly developed.

This trend for increased development and the changes that have taken place in the UK electricity industry make conflict between full land use and the presence of electricity infrastructure increasing likely. The aim of the book is to therefore give guidance in an area that is generally accepted to be highly technical. This is a practitioner's guide and so in this book we have focused on the dealings which an electric company has to have with an owner of private land, whether that owner is an individual or a corporate body. We are therefore treating the very particular considerations which arise from Crown lands or Church estates as outside the scope of this book. The dealings with which we are concerned, involving as they do both existing and new electric lines, are sufficiently controversial to fill a book on their own account. We may not be able to eliminate disputes entirely. However, we hope that with the benefit of this book such disputes as do arise can more efficiently be conducted and so more expeditiously resolved.

The law is stated at 1 April 2007.

Part 1

The Electricity Industry

Electricity Industry Structure

2.1 Outline of chapter

This chapter describes how the electricity industry developed so that we can more easily understand the differences between the old nationalised industry and the current privatised one; and, explaining why we must from now on talk about "licence-holders" and not "electricity companies". This chapter also illustrates who owns and runs what and in respect of which region.

2.2 From original legislation to nationalisation

The aim of the original legislation, the Electric Lighting Act 1882, was to "facilitate and regulate the supply of electricity for lighting and other purposes in Great Britain and Ireland". While this was primarily related to lighting this also provided the first definition of "electric line". Local distribution networks were established by private companies who also generated electricity and provided supplies to individual customers. In urban areas local authorities established electricity departments to carry out similar functions. A third category of joint electricity authorities completed the suite of authorised electricity undertakers. The rights obtained from owners and occupiers to run electric lines at voltages as high as 132,000 volts were the same as those that applied to local supply lines, then at 240 volts.

Change followed at a rapid pace in the early part of the 20th century. To meet the increasing demands for continuity of electricity supply the Central Electricity Board connected the regional networks of 132,000 volt overhead lines into a national gridiron. This was completed in 1933. This was not originally intended to be a national grid but a series of regional grids supplying local demand; they were however interlinked so that regions could help each other in supply emergencies. In 1938 they began operating as a national system.

The industry was nationalised in 1947 by the Electricity Act 1947. Under this Act the assets of the Central Electricity Board and the other authorised undertakers became vested in new statutory corporations. The 1947 Act also created the British Electricity Authority, renamed in 1954 as the Central Electricity Authority[1] which in 1957 became the Central Electricity Generating Board[2] (CEGB). The 1947 Act also created the 12 Area Electricity Boards of England and Wales.

So far as the general public was concerned the face of the electricity industry in the era of nationalisation would have been the Area Electricity Boards. This profile was supported by their high street presence in retail outlets, selling white goods as well as providing for bill collections. Local supply issues would also be dealt with by the Area Electricity Board. Apart from production of the commodity of electricity at power stations and transmitting it between power stations and grid sub stations, the Area Electricity Boards were the only real point of contact between the industry and the general public.

2.3 Further re-organisation

There was a further re-organisation within the nationalised electricity industry in 1972 relating to the level of voltage passing through an electric line. Those lines operating at 132,000 volts were the responsibility of the CEGB until 1972 but then became vested in the Area Electricity Boards. Until 1972 the public could reasonably have perceived all pylon lines as being the responsibility of the CEGB and, as will be explained in Chapter 3, this can make identification of the responsible organisation more difficult.

---

1   By virtue of The Electricity Reorganisation (Scotland) Act 1954.
2   By virtue of The Electricity Act 1957.

2.4 Privatisation

The statutory vehicle for privatisation was the Electricity Act 1989. The effective date for the transfer of property and assets to the privatised successors to the Area Electricity Boards was 31 March 1990 and the privatised industry was officially born on 1 April 1990. The legislative background and subsequent statutory developments are reviewed in Chapter 4.

The two main aims of privatisation of the electricity industry were to allow competition between suppliers of electricity and to allow new generators to connect to the system. These aims were prompted by the policy of reducing UK dependence on coal and fossil fuels and the increased availability of gas from the North Sea. No doubt with the life expectancy of many power stations approaching expiry there was an opportunity for private investment in what is inevitably a capital intensive industry. Privatisation was also intended to provide a better deal for customers in opening up this market to competition. Small service companies were able to enter the market; notably for metering and contracting services.

2.5 The emergence of the licence-holder

There are four stages in the supply of electricity to consumers: generation, transmission, distribution and supply. In England and Wales transmission is defined as being the transfer of electricity from the generating plant to a point of distribution at high voltage, which relates to systems operating in excess of 132,000 volts. At voltages below this value, the transfer of electricity between one point and another is known as distribution.

With effect from 1 April 1990 the four main elements of generation, transmission, distribution and supply were split between different companies with specific licences granted by the government and monitored by an appointed regulator.

The generation assets at that time were split between what became National Power and Powergen with the ability of new operators to enter the market.

Transmission assets in England and Wales became the domain of the National Grid Company plc.

England and Wales are divided into 12 distribution areas and responsibility in the respective areas rests with the Distribution Network Operator (DNO). The first DNOs were the Area Boards newly re-named as the local Regional Electricity Company (REC). As at 1 April 1990, therefore, the DNOs and the RECs were one and the same. This is illustrated diagrammatically at Appendix 1 and 2 and in the table at Appendix 3.

Supply was left with the former Area Electricity Boards, then becoming regional licence-holders under their new name of RECs with their licensed areas mirroring the original Area Electricity Board boundaries, as identified at Appendix 1.

Hence it is now necessary to refer to licence-holders and to avoid the term electricity company. We elaborate on the practical consequences below and give the relevant statutory definitions in Chapter 4.

The initial privatisation legislation permitted the same licence-holder to hold both a distribution and a supply licence at one and the same time. This is no longer the case as explained in more detail in Chapter 4.

2.6 Current industry structure

This book is essentially aimed at dealing with transmission and distribution lines across England and Wales. An illustration of how electricity is first generated and is then transmitted and distributed is given at Appendix 4. Distribution lines, being 132,000 volts and below, are the responsibility of the DNOs.

The vast majority of overhead tower lines operate at 132,000 volts on steel lattice towers similar in design but generally lower in height than towers used for the 275,000 and 400,000 volts overhead lines. These lines connect within sub stations to enable power flows to be moved around England and Wales to provide a safe, efficient and economic system. To give an indication of the range and scale of substation installations, within the West Midlands area, with a population of approximately 2 million, there are two sub stations operating at 400,000 volts, nine at 275,000 volts and forty at 132,000 volts.

The responsibility of the DNO ends at the cut out position which is normally immediately prior to the electricity meters. The meters themselves now provide critical evidence for identifying the licence-holder with which you are now actually dealing. In order to change supplier, the MPAN (Meter Point Administration Number) needs to be verified. The reading and verification of meters falls between DNOs and meter companies with a final billing system resting with a range of suppliers. It is the supplier who will pay the DNO for using their system and also pay the generators for the amount of electricity consumed. Understandably there is no direct single link between the generation of electricity and the individual consumer but the collecting and assessment of charges is carried out using the pooling arrangements.[3]

---

[3] The working of the pooling arrangement is beyond the scope of this book but further information can be obtained from the Association of Electricity Producers or the Department of Trade and Industry — Gas and Electricity Division.

Electricity Infrastructure

3.1 Types of electricity infrastructure

This chapter concentrates on the varying types of equipment used by the electricity industry expressed in layman's terms. Brief details will be provided on the following:

- statutory background
- steel lattice towers (pylons)
- alternative metal structures
- wood pole lines
- spacing between supports
- clearances
- tower and pole foundations
- underground cables.

3.2 Statutory background

The types of infrastructure used for electric lines in England and Wales are prescribed by a number of statutory instruments supplemented by detailed engineering recommendations and guidelines aimed at maintaining safety both to the public and to employees of and contractors engaged by licence-holders. The Act empowers the Secretary of State to make appropriate regulations. The Electricity

Supply Regulations 1988[1] cover these. Under these regulations the Department of Trade and Industry exercises control over standards through its engineering inspectorate to ensure compliance with the regulations and to report on incidents, particularly fatalities, which occur in relation to electric lines. The Electricity At Work Regulations 1989[2] provide for the statutory minimum clearances for overhead lines. The regulations governing clearance for overhead lines are supplemented by detailed guidance notes. These clearances vary in accordance with the normal use of the land over which the lines cross and therefore greater clearances are generally required for high speed road crossings and in connection with higher voltage lines. More detailed provisions relating to plant, overhead lines and underground cables are found in the Electricity Safety, Quality and Continuity Regulations 2002.[3]

The Health and Safety Executive also have detailed guidelines in their brochures GS6 — *Avoidance of Danger from Overhead Lines*,[4] HSG 47 — *Avoiding Danger from Underground Services*[5] and HSG 141 — *Electrical Safety on Construction Sites*.[6]

3.3 Types of overhead line support

Overhead line supports can be constructed of wood, metal or concrete. In other European Countries concrete structures are used but they are rarely encountered in England and Wales. Lower voltage lines, generally up to 33,000 volts, are predominantly supported on wood pole structures and higher voltage lines, including transmission lines, are generally on steel lattice towers, popularly referred to as pylons. Licence-holders adhere to the detailed requirements for the types of support required including the depth of foundations and clearance between conductors. They observe the common specification found in document ESI 43-8, last updated December 1988. This provides comprehensive data on the types of support required to maintain overhead lines of different voltages and in different situations.

---

1   SI 1988 No 1057.
2   SI 1989 No 0635.
3   SI 2002 No 2665 (subsequently updated by the Electricity Safety, Quality and Continuity (Amendment) Regulations 2006 (SI 2006 No 1521).
4   GS6 (3rd ed) HSE Books 1997 ISBN 0 7176 1348 8.
5   HSG47 HSE Books ISBN 0 7176 1744 0.
6   HSG141 HSE Books 1995. ISBN 0 7176 1000.

A general awareness of the limitations of the types of overhead line helps non specialised professionals to understand why the movement of single supports has a knock-on effect to the remainder of the line. Additionally, the apparently simple solution of replacing a section of overhead line with underground cable could prove to be significantly costly and have considerable implications aesthetically on the local environment.

3.4 Steel lattice towers (pylons)

Tower designs have changed in respect of their size and mass over the years, but otherwise are those in use since the early 20th century. The need for larger structures came about through the use of higher voltage lines which required heavier conductors and greater span distances.

Tower designs were reviewed by the National Grid Company in the early 1990s and a focus group studied the types of towers used in other European countries. The general consensus in terms of types of tower was that, for the most part, the types generally used produced the least effect on the environment, with careful siting. Furthermore, the colour finish, grey, is considered to blend in with the most predominant background created by the English weather! In addition to the voltage of the line the size and mass of a tower is also affected by the number of circuits that it carries and whether it is an angle tower. An angle tower is one where the line changes direction.

Most overhead lines carry two circuits, one on each side, with each circuit comprising three sets of phase conductors. Lines of 400,000 volts use quad conductors. Quad conductors are four individual conductors linked together by spacers for each phase which allows a large capacity of electricity to be transported across the country. Lines of 275,000 volts can also be of quad construction, but more commonly use twin conductors. At the next level down, 132,000 volts, it is also common for two circuits to be strung on tower lines with either single or twin conductors. Twin conductors are used for greater capacity circuits. In some instances only one circuit is carried and if the line were built to carry only one circuit then it would be likely to have two phase conductors on one side and one phase conductor on the other side. Illustrations of a typical suspension tower and also of typical high voltage towers in a 400kv route are given in Appendix 5.

In all cases overhead lines on steel towers carry an earth wire at the highest point on the tower. The earth wire, not used for transmitting or distributing energy is also capable of being the carrier for telecommunications apparatus. This is achieved by wrapping fibre optic cables around the earth wire, for which a separate wayleave is required if utilised for third party purposes. It is also possible for a separate catenary to be strung, purely for fibre optic purposes, and this is likely to be installed between the lower phases. A third alternative is to replace the earth wire with a fibre optic cable as the core of the replacement earth wire.

The design of towers is also affected by the need to provide electrical clearance between the conductors and the distance between the two circuits so that maintenance can be carried out on one circuit with the other circuit being maintained live. An illustration is also provided at Appendix 6 and 7 showing, in simple terms, what constitutes a phase carried on overhead lines and how two electrical circuits are carried on the same line. This diagram also gives a description of some of the components of an overhead line. A glossary of terms used in relation to electric lines is incorporated at Appendix 8.

3.5 Folded steel plate

A type of construction commonly used in parts of Western Europe is a folded steel plate design which produces a lower overall height than steel lattice towers with a reduced mass and is therefore more acceptable visually, particularly in built up areas. The downside to this type of construction is that it is only capable of carrying a single circuit. As with lattice towers it is also normally finished in grey. This type of tower has been used near business parks, notably the Cambridge Science Park, on the basis of its lower adverse impact on the local aesthetics of the built environment. This type of line would normally carry 132,000 volts.

3.6 Steel mast lines

Alternative supports utilised, predominantly for 33,000 and 66,000 volts lines, are steel masts which in appearance are comparable with vertical steel girders. This type of construction requires extensive use of stays at angle positions where a line changes direction. The benefit with this type of construction is that it can span greater distances and therefore require fewer supports when compared with wood pole lines.

3.7 P B structures

A limited number of structures were produced by Painter Brothers predominantly between the First and Second World War when wood poles were in short supply. These can be described as slim lattice towers with only one common column rather than towers having four separate legs. Again because of their rigid nature they require extensive stays at angle and terminal positions.

3.8 Wood Pole lines

The vast majority of electric lines in rural areas are on wood poles supporting two or three conductors carrying 11,000 volts. The poles are normally imported from Scandinavia and are pressure impregnated with creosote. A typical height of a pole would be approximately 10 to 12 m. As well as blending in with the countryside, wood poles facilitate access to the conductors by linesmen who can climb the poles using specially designed boots with spikes and applying a harness to their upper body to allow manual work at the crossarms.

3.9 Trident Construction

An alternative to a tower line is to utilise wood pole lines for carrying 132,000 volts. Such lines utilise a separate specification known as trident construction, due to its upward pointing insulation resembling the Fork (or Trident) of Neptune (or Poseidon depending on whether your preferences are for Greek or Roman mythology!). These are only capable of carrying a single circuit but as they are of a lower profile they have a reduced overall effect on the environment. The weight of the conductors, being greater than typical similar lines at 11,000 and 33,000 volts, accentuates heavy angles and requires the use of stout poles and in many cases double pole arrangements.

3.10 Spacing between supports

The space between different supports depends on the type of support and, while there is a maximum span distance, there is also a need to maintain the tension of the line and therefore the balance of the distance over two spans is generally taken into account when calculating the maximum distance, always ensuring that satisfactory ground clearance is maintained in accordance with the normal operations of the land.

Typical spacing between towers for most transmission lines is approximately 300 m. In some cases single span distances can be greater than this but on a typical length of line this would be the average span between towers. A notable exception to this is the crossing of the River Severn immediately south of the Severn Bridge with a single span across the Bristol Channel. Most lines at 132,000 volts use a standard specification known as L4M with the larger towers using L2, L7 or L12. Each design is specific in its nature and where greater clearance is required to ground level, and where crossing existing buildings, then extensions are incorporated in the basic tower design in 3 m increments. With wood pole lines a typical span distance would be 100 m with greater distances provided where double poles, referred to as H poles, are utilised.

3.11 Clearance to buildings and obstacles

All electricity lines are designed to provide for a safe working clearance from ground levels and buildings when the line is originally built. The requirements for spatial clearances are contained in a comprehensive document prepared by the electricity industry and known as ESI Standard 43-8, last issued in December 1988. As well as identifying clearances at all system voltages from general obstacles, it also includes clearances for specific areas, notably operational railway lines and watercourses. Clearances have been designed to provide safety to the general public and protection against flashover of the line. The higher the voltage of the line the greater the flashover distance generally. An electric shock is possible by the current jumping from the conductors to a person or conductive object as it is not merely a case of making physical contact with live conductors. As identified in the standards document only general guidance can be given and in specific locations it would be appropriate to communicate with the relevant licence-holder to determine safe working clearances and safe working arrangements. Additional considerations may be required where induced voltage may occur, for example where metal fences are planned to run parallel with high voltage lines. Typical clearance in areas where vehicles could pass beneath would be 5.8 to 6.7 m for lines designed to operate at less than 132,000 volts and a minimum of 6.7 m for lines of 132,000 volts increasing to 7 and 7.3 m for lines of 275,000 and 400,000 volts respectively. Clearances to trees and objects on which a person can stand are provided at Appendix 9 and 10 respectively.

3.12 Tower foundations

Given the significant height of towers relative to their base the stability required is critical. As a result the foundations, typically concrete, need to be based on firm sub strata and may well extend to 5 or 7 m below the ground surface. The foundations themselves can be compared to a pyramid structure and offer sufficient mass to avoid movement. Suspension towers would typically have four equal foundations. At angle towers, two legs of the tower would have stresses inward from the angle of the line and need to allow for the compression effect of the tower. Alternatively the two external legs of the tower also need to allow for the uplifting effect generated by the force of the conductors on the tower and the engineering specification needs to allow for this. Where erosion of the immediately adjacent ground area is possible, for example adjacent to watercourses, then additional foundations would be installed to allow for this. In existing situations it has also been known for Gabion Walls to be constructed, running parallel with the watercourse, to protect the tower foundations from erosion.

3.13 Pole foundations

In comparison to the installation of telecommunication poles, often referred to as telegraph poles, where a vertical borer or auger is typically used, excavations for wooden poles for electric lines are normally excavated on one side only to allow the pole to be pulled into place with a firm base on at least 180° of unaffected ground. In further comparison to telecommunication poles the poles used for electric lines require greater stoutness and depth and are typically installed at a depth of approximately 2 m. In areas of poor ground condition, such as peat or bog conditions, they would require timber cross members to be installed known as bog shoes. Metal stay wires, to support angles where lines change direction, are installed with a stay block in the ground, comparable to a railway sleeper, at 90° to the pole.

3.14 Underground cables

In urban areas underground cables form the bulk of connections for the electricity distribution system. 230 volts applies generally for single domestic supplies, and 415 volts for three phase supplies, largely for industrial uses. Underground cables can be installed with a capacity for distributing up to 400,000 volts. The expression cable is used to describe underground electricity cables whereas the term line is applied to overhead electricity lines. Most cables are installed directly into the ground and in a new housing development would be laid in service strips or beneath footpaths in accordance with the agreed convention in relation to other underground utility apparatus. The depths at which cables are laid generally increases with the voltage with the aim being to provide a sufficient depth of cover over the cables so as to minimise potential contact by third parties. At higher voltages cables are more likely to be installed in ducts and in all cases would be surrounded by sand to avoid the impact of bricks and stones damaging the cable when the trench is backfilled. The depth required for cables of 132,000 volts would generally fall in the range of 1–1.2 m with sufficient spacing required between separate cables. This is both to minimise the effect of dissipating heat influencing other cables and to allow maintenance work or repairs to be carried out on discrete circuits. Underground cables at 132,000, 275,000 and 400,000 volts would additionally require spacing between separate cable circuits to minimise the impact on adjacent circuits if one circuit were to fault in close proximity to other cables. The depth of cover above cables is a critical factor in calculating the extent of heat dissipation as excessive depths could impact on the cable rating and potentially necessitate an increase in cable sizes.

The restrictions on development in proximity to cables are influenced by the rights that have been granted. Where a wayleave has been granted it is likely to specify the number of cables to be installed and possibly the operating voltage, but not a specific working width. Where an easement has been granted then the width of the land within the servient tenement[7] affected by the easement is usually specified and would require that no building is constructed over the easement route nor trees planted within the easement strip. Additionally there may be restrictions on the planting of trees in proximity to the

---

7   See section 5.11.2 for an explanation of this term.

easement route where the roots may impact upon the operation of the underground cables. This is covered in greater detail in Chapter 15.[8]

On older operating systems oil filled cables were utilised for higher voltages. Non-fluid filled cables are now generally utilised and are protected by insulating membranes. In some cases the cables will be installed in plastic ducts; both to provide additional protection and for ease of replacement. In addition to installing underground cables for the distribution or transmission of electricity, pilot or communication cables may also be laid in the same trench to facilitate the operations of the licence-holder for electricity purposes.

---

8   See in particular section 15.14.

Energy Legislation Since 1989

4.1 Outline of chapter

From the point of view of this work, the Electricity Act 1989 (the Act) remains our principal text. However, the Act has been substantially amended since it was first passed. Accordingly to guide us through this book we not only need an overview of the Act as it currently exists but we also need to be aware of how the Act fits into the general context of energy legislation since 1989. As will be seen we are concerned with a small and carefully defined part of the statutory picture but the landscape in the background must be understood if mistakes are to be avoided.

4.2 Overview of statutory history since 1989

The purpose of the Act was to reorganise the electricity industry completely. The industry had been nationalised by the Electricity Act 1947. By the Act the industry was made ready for privatisation. Accordingly the Electricity Act 1947 was repealed in its entirety along with numerous other acts and statutory provisions which had presupposed and cross-referred to the previously nationalised industry.[1] In its place Part 1 of the Act dealt with licensing, the duties

---

1   Electricity Act 1989, Schedule 18 which lists 90 earlier statutes or statutory provisions.

of those engaged in electrical distribution, enforcement of licence conditions and consumer protection. We continue to be concerned with certain of the provisions of Part 1 as they now stand. Part 2 dealt with the transfer of property rights and liabilities and the financing of successor companies. Under section 65 of the Act, the Secretary of State was obliged to nominate a company to take over as at the appointed date all property rights and liabilities to which each Area Board was entitled or subject and the transfer was to take effect on the date nominated by him. Under section 66 of the Act, the Generating Board and the Electricity Council had to make a scheme for the division of their respective property rights and liabilities and on the appointed date, transfer to any company nominated by the Secretary of State as a transferee under that scheme those property rights and liabilities. The appointed date was 31 March 1990. Part 3 of the Act was the Miscellaneous and Supplemental Part and as the heading suggests this was the part in which a variety of consequential matters are addressed. The principal section of continuing relevance to this work is section 109 which deals with formal service of documents.

The Competition and Service (Utilities) Act 1992 made provision for standards of performance and service to its customers in relation to all utilities. The result was that the provisions of sections 40(3), 42A and 44A were inserted. These provided a structure for industrial regulation but have been yet further modified. The provisions for the regulation of competition were further amended by the Competition Act 1998 and the Enterprise Act 2002.

However there was far wider reform as a result of the Utilities Act 2000. According to the pre-amble of the statute, the purpose of the Utilities Act 2000 was "to provide for the establishment and functions of the Gas and Electricity Markets Authority, ('the Authority') and the Gas and Electricity Consumer Council; to amend the legislation regulating the gas and electricity industries; and for connected purposes."

Legislation relating to the reduction of carbon emissions and the government's energy policy generally are to be found in the Sustainable Energy Act 2003 and the Energy Act 2004. With limited exceptions the provisions of those statutes are outside the scope of this book.

The Energy Act 2004 while principally concerned with the nuclear industry and renewable energy resources, also inserted a number of additional provisions into the Act particularly as regards public inquiries. It also added a definition of "high voltage lines" and "low voltage lines".

4.3 Current industry regulation

The Authority is now the industry regulator. Section 3A of the Act states that the principal objective of the Secretary of State and the Authority is to protect the interests of consumers in relation to electricity conveyed by distribution systems. It grants licences to companies[2] without which it is a criminal offence[3] to generate, distribute or supply electricity or participate in the transmission of electricity for the purpose of supply. The Authority regulates the conditions[4] under which such licences are held, controls the transfer of licences,[5] refers any matter to the Competition Commission[6] and secures compliance with conditions of licences.[7]

The role of the Authority is to ensure fair competition and to provide, through five yearly reviews, a review mechanism of the pricing structure imposed by licence-holders. Matters relating to wayleaves ought not to be referred to the Authority as this goes beyond its terms of reference. A wayleave is considered to be a separate contract between the relevant licence-holder and the landowner, who will not be acting in his role as customer but in respect of his ownership or occupation of land. Matters of inefficient administration can however be referred to the Authority who also maintains a register of complaints against individual licence-holders as one means of assessing their efficiency.

The public face of the Authority is the Office of Gas and Electricity Markets more commonly know by its acronym Ofgem. The Ofgem website[8] describes Ofgem as "being governed by the Authority". Strictly it is the Authority but in its executive capacity. What this convenient shorthand means is that the Chief Executive of Ofgem and Ofgem's executive directors run the Office on a day-to-day basis. The Authority also has members who are non-executive. Full details are likewise available from the web-site. The executive and the non-executive members meet formally as the Authority to plan strategy amongst other things. The Authority has published a Corporate Plan for the period 2006–2011. It is a comprehensive document which is

---

2 Section 6 of the Act.
3 Section 4 of the Act.
4 Section 7A of the Act.
5 Sections 8A, 11 and 11A of the Act.
6 Section 12 of the Act.
7 Section 25 of the Act.
8 *www.ofgem.gov.uk*.

available from the website. The issues raised are beyond the scope of this book. Suffice it to say that anyone concerned with the electricity industry who is seeking to evaluate current trends and plan for future challenges would be very well advised to read this corporate plan.

4.4 Current licensing regime

The Utilities Act 2000 did not just establish the Authority. It created the current licensing regime over which the Authority presides. In consequence the Competition and Service (Utilities Act) 1992 was repealed. As indicated in Chapter 2 we now have the following classes of licence which a licence-holder may obtain:

- a generation licence (section 6(1)(a))
- a transmission licence (section 6(1)(b))
- a distribution licence (section 6(1)(c))
- a supply licence (section 6(1)(d))
- an interconnector licence (section 6(1)(e)).

A supply licence is for those who have the most immediate contact with the consumer. An interconnector licence is concerned with networks between sovereign states. Those holding a generation licence may need a connection to the transmission or distribution network. If that connection in turn requires wayleaves across private land, the obligation to connect generating plant to the system falls to the transmission or distribution licence-holder who would then seek to obtain all wayleaves. It is the holders of transmission licences and distribution licences with which we are principally concerned in this book. They are the licence-holders who need non-statutory wayleaves or the facilities offered by the statutory menu[9] if they are to discharge their obligations.

What a licence-holder may not do is hold both a supply and a distribution licence at the same time. Accordingly the Utilities Act 2000 established a framework for the transfer of property and liabilities to associate companies to enable licence-holders to comply. The effective date for such transfers was 1 October 2001. For this reason when considering which licence-holder has the benefit of a non-statutory

---

9   See Chapter 8.

wayleave, that is the crucial date to which the questions discussed in Chapter 7 need to be directed.

A list of all current licence-holders is available from the Ofgem website.

4.5 Relevant provisions of the Act as we now have it

As explained above, the Act as amended by subsequent legislation is divided into 3 parts and has schedules 3, 4, 6–12 inclusive, 14, 16 and 17 annexed to it. The schedules with numbers outside that sequence have either been repealed or apply to Scotland only and so are outside the scope of this work.

Section 10 of the Act deals with powers of licence-holders which include compulsory acquisition. By section 10(1) of the Act, schedules 3 and 4 to the Act are made relevant to the exercise of those powers. Schedule 3 relates to Easements and is considered in Chapter 15. Schedule 4 relates to wayleaves and ancillary rights and is considered extensively in Part 3 of this book.

Sections 36 and 37 of the Act deal with consents from the Secretary of State for the construction or expansion of a generating station (section 36) or the installation of an electric line above ground (section 37). By section 36(8), schedule 8 to the Act is made relevant to an application for any such consent. Section 62 gives the Secretary of State power to order a public inquiry subject to the terms of reference imposed by that section. The issues raised by those provisions insofar as they are relevant to our theme of maintaining and installing electric lines on private land[10] are considered in Part 3 of this book. The interactions of requiring consent under section 37 and wayleaves to be granted by the Secretary of State under Schedule 4 are considered in Chapter 8.[11] Any proposal requiring consent under sections 36 or 37 also requires environmental aspects to be considered by virtue of schedule 9 to the Act. These obligations are considered further in this book.[12]

---

10   Amendments made by the Energy Act 2004 relating, for instance, to off-shore generating stations are outside the scope of this book.
11   See the overview in Chapter 8, section 8.7 and the further cross-references there.
12   See Chapter 8, sections 8.8 and 8.9.

4.6 The Secretary of State

References have been made to the Secretary of State. By section 5 of schedule 1 to the Interpretation Act 1978, the Secretary of State means one of Her Majesty's Principal Secretaries of State. In theory any of the Principal Secretaries of State can carry out any duty remitted to the Secretary of State. In practice the work of government is divided among recognised departments with distinct responsibilities. For the purposes of this book, the Secretary of State means the Secretary of State for Trade and Industry. Applications for Consents under sections 36 and 37 and Schedule 8 and necessary wayleaves under section 10 and Schedule 4 to the Act are managed by the Department of Trade and Industry.

4.7 Electric line

An electric line means any line which is used for carrying electricity for any purpose and includes, unless the context otherwise requires:

- any support for any such line, that is to say any structure, pole or other thing in, on, by or from which any such line is or may be supported, carried or suspended
- any apparatus connected to any such line for the purpose of carrying electricity and
- any wire, cable, tube, pipe or other similar thing (including its casing or coating) which surrounds or supports, or is surrounded or supported by, or is installed in close proximity to or is supported, carried or suspended in association with any such line.[13]

An electric line is a convenient shorthand for the apparatus which is the subject-matter of a wayleave, easement or consent and so that is the term we shall use throughout this book.

---

13   Section 64(1) of the Act.

Part 2

Private Rights

Non-Statutory Permissions Granted by Private Agreement

5.1 Introduction: What non-statutory permissions are the subject-matter of this chapter?

There are five categories of non-statutory permissions which we need to consider. The first two are the primary areas where permissions are required and the following three are complementary to this. Each of these areas will be considered in turn:

- wayleaves
- easements
- rights supplementary to a wayleave without which the wayleave is less effective (ancillary rights)
- rights to lop and fell trees in order to facilitate the exercise of the wayleave
- rights to enter land for surveying.

5.2 Outline of chapter

This chapter gives a historical introduction to the development of the concept of a wayleave. There is no comprehensive definition of a wayleave either in any of the cases decided by the courts on wayleaves or in any of the relevant legislation. This chapter, therefore, considers whether there are sufficiently clear indications from statute and the cases for us to be able to develop a working definition which will enable

33

us to identify all forms of non-statutory wayleave and distinguish those from a "necessary wayleave"[1] granted under statutory powers. We also compare wayleaves with other legal forms, notably easements, to prepare the ground for the discussion in Chapter 15. It also comments upon ancillary rights including the lopping and felling of trees and rights to survey land.

5.3 Why are wayleaves needed in the first place?

The wayleave is our unit of legal currency in the market in which we are now operating. We therefore need to understand: why we need it and what benefits it confers on those to whom wayleaves are granted. We also need to be able to distinguish a wayleave from an easement or a lease or tenancy.

It is essential to ensure both that the installation of an electric line and that entry on to land to inspect, repair and maintain it once installed are covered by permissions granted by a wayleave.

In relation to *entry* upon land a wayleave prevents the person entering from trespassing upon the land. The permission makes the entry lawful.[2] Trespass is not, though, confined to single or occasional or even frequent and repeated entries. "Trespass to land consists in any unjustifiable intrusion by one person upon land in possession".[3] So the use of a person's land for the laying of an electric line would be a trespass in two respects: the original entry in connection with the works of installation and the continuous presence of the line on the land. If the non-permitted use of the land continued for 12 years and the licence-holder went through the procedures laid down by the Land Registration Act 2002 then in theory the licence-holder might be able to claim prescriptive title by adverse possession to the land occupied by the electricity line. However, that would not be the same as an easement. As discussed below, for an easement you need a dominant landholding which the right accommodates. Moreover the prescription period for an easement is 20 years. Not only that: if the claim is that you have had exclusive and unrestricted possession of

---

1   Considered in Chapters 8–10 inclusive.
2   *Clerk and Lindsell on Torts* 19th ed, paragraphs 19–45.
3   *Clerk and Lindsell on Torts* 19th ed, paragraphs 19–01 and the authorities there cited.

land that eliminates a claim for an easement in any event.[4] Possession of the land on which the electric line is located does not on its own give a right to enter and repair and cross over other land in order to access the electric line for the purpose of inspection, maintenance and repair. For that you need a separate licence or an easement.

5.4 Wayleaves: historical background

All major legal developments are responsive to significant economic changes or social restructuring. The law of wayleaves is no exception. A branch of the law which began life as an answer to the question: "can I walk over or drive my cart or cattle over your land?" and remained quietly in its humble place for many years was thrust into the celebrity spotlight in the mid-19th century.

A wayleave was first mentioned in a law report in 1840. The case[5] concerned a deed of 1630. By the deed, land in the manor of Amble, Northumberland, had been conveyed from one person (the grantor) to another person (the grantee) subject to the reservation of a right in favour of the grantor of "all mines of coal within fields and territories of Amble together with sufficient wayleave and stayleave to and from the said mines and the liberty of sinking and digging pit and pits".

The question in the case was whether this reservation of a "sufficient wayleave" gave the grantor as a coal-owner the right to use an existing railway for the purposes of carrying coals from Amble to Hauxley (also in Northumberland) and *vice versa*. The crucial fact was that if that particular use were permissible, the grantor as coal-owner would be using his wayleave and stayleave granted by a deed relating to land at Amble not in connection with coals mined in Amble but in connection with coals mined in Hauxley. Even though as a matter of geology, coal both in Amble and in Hauxley was within the same minefield, was the grantor as coal-owner trying to use the wayleave and stayleave more intensely than originally intended and for a purpose different from that originally contemplated by the parties? The grantee certainly complained that that was the case. The grantee claimed that the grantor was abusing his reserved right by doing more

---

4 Hanina v Marland [2000] 97 (47) LS Gaz.41 and see further the discussion *Clerk and Lindsell on Torts* 19th ed at paragraphs 19–42.
5 Dand v Kingscote 6 M&W 174 or [1835–42] All ER 682.

than was permitted and so the grantor was liable to compensate the grantee for wrongful use of the grantee's land.

Counsel for the grantee defined "wayleave and stayleave" as follows:

> the term stayleave according to custom and the understanding of miners and other persons conversant with coal mines means a right in the coal-owner of having a station where he may deposit his coals for the purposes of dispensing them to the purchaser. Wayleave is the privilege of crossing land for the supply of coal to the purchaser. This privilege is generally the subject of a detailed contract specifying the particular direction and extent of the wayleave and there is no usage of or understanding amongst persons conversant with coal mines by which to interpret the extent of the privilege when conferred in general terms (such as those used in the 1630 deed).

The court in considering the reservation held that "sufficient wayleave and stayleave" meant

> a right to such a description of wayleave and in such a direction as will be reasonably sufficient to enable the coal-owner from time to time to mine all the seams of coal to a reasonable profit and therefore the owner is not confined to such description of way as in use at the time of the grant and in such direction as is then convenient.

So the question became whether the direction or mode of construction of the railway was reasonably sufficient for the purpose of getting coal from the Amble coal mine. The ruling was that the grantor's coal-owner had gone too far and so was liable in damages. The remainder of the case considered the basis on which damages should be assessed.

5.5 Survival of the concept of a wayleave into the era of privatisation

The term wayleave was adopted by the electricity industry because the right to string electric lines on private land was considered less onerous than a lease or easement and did not require total ownership of land. Other utilities obtained stronger and more intrusive rights to support the growth of industry in the 19th Century — essentially for transportation and to improve public health. In order to access poles or towers (pylons) a permission or "leave" to make one's way over land, as opposed to the more formal access rights granted for the construction of public highways and sections of railway, was considered sufficient for this purpose.

5.6 Estates and interests in land: what is "land"?

Before considering what hallmarks the law has held a wayleave to have, it is important to understand first what the law means when the word land is used and, second, to understand the hierarchy of land rights which the law recognises.

Section 205(1)(ix) of the Law of Property Act 1925 provides that whenever the word "land" is used in that Act it includes:

> land of any tenure, and mines and minerals, whether or not held apart from the surface, buildings or parts of buildings (whether the division is horizontal, vertical or made in any other way) and other corporeal hereditaments; also a manor, an advowson and a rent and other incorporeal hereditaments, and an easement, right, privilege, or benefit in, over, or derived from land.[6]

More generally, by virtue of section 5 of the Interpretation Act 1978, unless any Act contains a specific definition of land such as the one which appears above the word land carries the definition contained in schedule 1 to the Interpretation Act 1978 namely land "includes buildings and other structures, land covered with water, and any estate, interest, servitude or right in or over land."

5.7 Estates and interests in land: what is the hierarchy?

At the top of the hierarchy are legal estates and interests. Below them are equitable estates and interests. Below them are rights to use land which fall short of equitable estates or interests. Let us elaborate on each category in turn, bearing in mind that the relevant definition of land for these purposes is the one given by the Law of Property Act 1925.

By section 5(1) of the Law of Property Act 1925:

> The only estates in land which are capable of subsisting or of being conveyed or created at law are —

---

[6] For a fuller discussion of this topic and in particular the technical terms used in this and the next quotation, the reader is referred to *Megarry & Wade: The Law of Real Property* 6th ed, edited by Dr C Harpum of Falcon Chambers, London or to *Cheshire and Burn's Law of Property* 16th ed.

- An estate in fee simple absolute in possession
- A term of years absolute

The only interests or charges in or over land which are capable of subsisting or of being conveyed or created at law are an easement, right or privilege in or over land for an interest equivalent to an estate in fee simple absolute in possession or a term of years absolute.

"Fee simple absolute in possession" means a freehold interest. A freehold interest lasts for ever and gives the freeholder the fullest possible rights to deal with the land. "A term of years absolute" means a lease. It is a term because the right to possess and deal with the land is carved out of the freehold interest[7] for a fixed period. These estates and interests bind the land and can therefore survive changes in ownership. Their key feature is that the rights attached to such estates and interests are enforceable against any third party whatsoever by the person in whom they are formally vested.[8]

The only way a right in respect of land could amount to a legal estate or a legal interest is if it falls into one or other of the above categories. In other words for an easement right or privilege to be granted as a legal interest in land it must either be granted by a deed granting a permanent right in which case, just as a freehold interest has not, it has no end date; or it must be granted as an integral part of a lease in which case it will last for as long as the lease does.

An equitable estate or interest in land is one which carries with it a right to occupy or deal with the land which may or may not be limited in time or extent and which may fall short of "legal estate or interest" status because of some defect in the document by which it was granted or lack of proper registration. They may be equitable rather than legal because they entitle the person with the benefit of the estate or interest to the proceeds of sale of the land rather than the possession of the land itself. An equitable estate or interest is vulnerable to the risk of a party claiming that it has come to an end because a right higher up the hierarchy takes precedence. These estates and interests while clearly

---

[7] For the purposes of this discussion we have ignored interests superior to the tenant in possession which are nevertheless themselves leasehold. For a fuller discussion the reader is referred to *Megarry & Wade: The Law of Real Property* 6th ed, edited by Dr C Harpum of Falcon Chambers, London or to *Cheshire and Burn's Law of Real Property*, 16th ed..

[8] In respect of registered titles this will mean the person registered at the Land Registry as the proprietor of the estate or interest.

related to land nevertheless do not carry with them the ability to enforce the rights attached to them against all third parties. They often depend on first the ability to enforce a contract against a particular third party and then the ability to prevent that particular third party from acting contrary to that contract.

5.8 What are the recognised practical hallmarks of a wayleave?

There is within the electricity industry a commonly accepted and conventional understanding of what a wayleave is. A wayleave is a permission granted by an owner or occupier of land which enables a third party to cross land with equipment or vehicles in connection with some business venture or to lay in or under land or install upon it some infrastructure.

5.9 Legal features common to all non-statutory wayleaves[9]

How does this understanding measure up against a strict legal analysis? Whether express or implied,[10] all wayleaves will have the following features: a permission to install an electric line on defined land is granted by a landowner or occupier to a licence-holder in return for payment of a fee — the wayleave fee.[11]

Such features amount to an offer (namely to pay for the privilege of using defined land for a specific purpose), an acceptance (in the form of the grant of permission) and the passing of consideration (namely the actual payment of the wayleave fee).

Offer, acceptance and the passing of consideration in the context of an arrangement designed to create legal relations are the essential hallmarks of any contract.

---

9   See Chapters 6 and 7 for a fuller discussion of the specific characteristics of wayleaves in this category.
10   See Chapter 6 for express and Chapter 7 for implied wayleaves.
11   Such features are also, of course, exhibited by a necessary wayleave — see Chapter 8.

It follows that all non-statutory wayleaves[12] are by definition contractual. That would not necessarily be the generally accepted view. Contract usually means in common parlance a formal legal document. As will be explained below, that is not legally correct.[13] The consequences of the fact that all non-statutory wayleaves are contractual are first that there will not automatically be terms over and above the bare essentials and secondly that the wayleave can only be brought to an end in accordance with its terms.

A right to use land which falls short of any of the categories in section 5.6 is one which does not carry with it any degree of permanence; or does not carry with it any ability to deal with or occupy the land; or can be terminated at the discretion of the grantor of the right. Such rights tend to be informally granted and are not capable of binding successors in title. Wayleaves are rights which though contractual need not be granted in writing and can arise by implication.[14] Further as explained below they are rarely if ever granted by deed. In consequence a wayleave could never be a legal estate or a legal interest in land.

It is noteworthy that the necessary wayleave procedures (see Chapters 9 and 10) presuppose that wayleaves terminate on a change of ownership on the part of the grantor[15] which is consistent with the status of wayleaves as being that of a contractual licence and therefore conferring no interest in the land in respect of which it is granted.[16]

In so far as wayleaves have been considered by the courts, since *Dand v Kingscote* (6 M&W 174 or [1835–42] All ER 682), the focus has been not so much on the property aspects as on different issues. The reason why the property and ownership aspects of wayleaves received minimal attention is that the existence and use of wayleaves were taken as read. The question before the courts when wayleaves were considered was how you would deal with compensation for wrongful use of land. We go into that question in more detail in Chapter 21. This may be explained by the fact that if wayleaves were temporary rights only, they would not attract much judicial attention. Being mere

---

12   The word "voluntary" which is in common use is considered in Chapter 6, section 6.3.
13   For a full discussion see Chapter 6.
14   See Chapters 6 and 7.
15   See Chapter 6, section 6.6 for an explanation of "grantor".
16   As discussed in Chapters 4 and 7, the same result does not occur on a change of licence-holder.

precarious privileges they would cause few difficulties. Judges can only give their opinion on the cases brought before them and so if wayleaves were not producing litigation, they would not feature in the case-law. There is some guidance, however, in the case of the *North Eastern Railway Co v Lord Hastings* [1900] AC 260. This concerned a wayleave granted by a document whereby a railway was permitted to cross land. For our purposes it emphasises the crucial importance of the wayleave depending on the precise terms of the document but it did not analyse any of the property aspects mentioned above.

Apart from a passing reference in 1908[17] wayleaves as such did not trouble the court again until 2003[18] where the issue was the correct understanding of the word "land" under paragraphs 6 and 7 of Schedule 4 to the Act. The case will therefore be discussed in Chapter 8. However, the *British Waterways* case did not deal with any of the questions now under consideration.

The only other areas considered by the courts were the extent to which statutory powers to lay pipes and railways carried with them an implied right of support and are therefore outside the scope of this book. However in considering what might be the property and ownership aspects of wayleaves, some guidance may be drawn from a decision that a statutory right to lay pipes conferred only a right of exclusive occupation and not a tenancy or an easement.[19]

5.10 Wayleaves as a legal category

To summarise the preceding discussion the only clues the cases give us as to the legal hallmarks of a wayleave are the following:

- a wayleave is a privilege granted by permission of a landowner or occupier
- it relates to use of land in connection with a specified purpose rather than ownership of the land
- the permission can be limited by terms
- if the boundaries of the permission are exceeded then compensation will be payable.

---

17 *New Moss Colliery Co Ltd v Manchester Corporation* [1908] AC 117.
18 *British Waterways Board v London Power Networks plc* [2003] 1 All ER 187.
19 *Newcastle-under-Lyme Corporation v Wolstanton Ltd* [1947] 1 Ch 427. The case should be treated with caution since it was a decision on a specific statutory provision.

The legal category into which a wayleave most neatly fits in those circumstances is a contractual licence.[20]

A licence as discussed is a mere permission which makes it lawful for the person to whom the licence is granted to that which would otherwise be wrongful. In the context of the statutory regulation of the electricity industry we talk about the electricity provider being a licence-holder. In the context of land law, we describe the person giving permission to use his land for any purpose as the licensor and we describe the person receiving the permission as the licensee. A licence converts what would otherwise be a trespass to land into a lawful act.

The effect of the licence being contractual is the rights and liabilities of the licensor and the licensee, the period for which the licence continues and the provision for revoking the licence are all governed by the agreement made between the licensor and the licensee. That agreement is a contract. The licence will not be validly revoked unless the contractual provisions are followed. Usually the provision is that notice has to be given in writing so many months in advance of the intended termination date.

A contractual licence does not create any interest in land.

The hallmarks of a wayleave all fit this category. An agreement to permit use of the land (however informally recorded or made) in consideration of the payment of a wayleave fee is a contract. A wayleave is not independently recognised as creating an estate or interest in land.

To corroborate the suggestion that a wayleave is best explained as a contractual licence we should compare and contrast a wayleave with both a tenancy and an easement.

5.11 Wayleaves contrasted with other legal forms

5.11.1 Tenancies

A tenancy gives an interest in land for a specified period or until a specified notice be given. The interest is given subject to conditions and obligations (covenants). A tenancy granted by deed is a lease. An interest in land can in principle be sub-let or assigned. A personal right to cross land cannot.

---

20 See *Ashburn Anstalt* v *Arnold* [1989] Ch 1 and the fuller discussion in the leading text books on land law: *Cheshire & Burns* 16th ed and *Megarry & Wade* 6th ed.

5.11.2 Easements

An easement is a right such as a right of way or a right of light or a right of support which one property concedes to another. The property which concedes the right is called the servient tenement. The property which enjoys the right is called the dominant tenement. An easement binds both the land granting the right and is attached to the land benefiting from the right and so it not only persists through successive ownerships, it is also capable of being registered as an adjunct to a title.

At[21] common law an easement can only be granted expressly by deed, usually by express[22] grant or reservation contained in a conveyance or transfer. Easements are also frequently contained in leases. However, separate deeds of easement can be executed. The grant must make it clear that it is granting an easement. A deed stating that the grantors "hereby grant their licence and consent" will create only a licence, which cannot pass an interest in land even though the grant is expressed to be for the grantors' successors in title.[23] The reader should also be aware that there are additional complex requirements relating to the registration of easements created in relation to registered land.[24] The scope of easements, their terms and valuation aspects are considered in Chapter 15.

To be the beneficiary of a wayleave you do not have to have land which is accommodated or served by the wayleave. As the beneficiary of the wayleave you do not have to have a related land interest at all. Your infrastructure may be entirely free standing and independent of your own particular land holdings. Nor does the beneficiary need to own the land on which the infrastructure stands.

Furthermore, in contrast to the formalities associated with easements, wayleaves can be (and very often are) created without a deed. We discuss this in more detail in Chapters 6 and 7.

In consequence a wayleave could hardly ever qualify as an easement and even if it were in fact granted as such, to qualify as a legal estate or a legal interest in land the wayleave would have to be granted on terms and in a manner which meets the formal requirements discussed above. As explained in Chapter 15, easements

---

21   The text in this paragraph is a quotation taken from *Boundaries and Easements* by Colin Sara (3rd ed) and we express our thanks.
22   For a discussion of the legal definition of express, please see Chapter 6, section 6.5.
23   *IDC v Clark* (1993) 65 P&CR 179.
24   For a full discussion, the reader is referred to Mr Sara's book already mentioned.

are from time to time granted voluntarily but non-statutory wayleaves are far more common. In fact, during the period of the Area Electricity Boards, the granting of wayleaves was made significantly easier by the development of terms which were substantially standardised and which were incorporated into voluntary wayleaves.[25]

For similar reasons, tenancies never carried popular support as vehicles for rights to install electric lines presumably because it would have been necessary for the grantor to concede possession of the land used and also because the creation of tenancies and leases would often involve more time and formality.

Finally, we must say a word about registration of title. A traditional wayleave being granted not by deed as a legal estate or a legal interest cannot be registered at the Land Registry. However, because it is neither a legal estate nor a legal interest nor an equitable estate nor an equitable interest, it does not survive a change of ownership in the land over which the wayleave is granted either. The implications of this are discussed in Chapter 7. The point is a wayleave will not be discovered by looking at a register of title.

5.12 Conclusions: a working legal definition of a wayleave

For the purposes of this book we can define a wayleave as a permission enabling a person to lay or install infrastructure on or over land belonging to or in the occupation of another and which permission includes the ability to enter the land to inspect and maintain the infrastructure. A wayleave creates no interest in land.

A wayleave is usually granted in the form of a contractual licence either for a fixed term or subject to the right on the part of the grantor to terminate it upon notice. The licence may additionally impose such other terms as the grantor and the licence-holder negotiate. As we shall see in Chapters 6 and 7 the contractual licence may be an express or an implied agreement.

Being a contractual licence, apart from any provision relating to termination included in the licence itself, the wayleave automatically terminates in the following circumstances:

---

25 See Chapter 6, section 6.8.

- if the grantor ceases to own the land over which the wayleave is granted
- the grantor permits another to occupy the land with a superior estate or interest
- or the grantor dies.

There is no central wayleaves register. Whether there ought to be one and if so who should maintain it is a question we consider in Chapter 22. Wayleaves are not registrable interests for the purposes of the Land Registration Act 2002. It is not possible to ascertain whether any wayleave exists except by factual enquiry. Therefore you have to understand the legal features of a wayleave if you are faced with circumstances which require you to find an answer to the question: do we have one? Furthermore, you have to understand the restrictions on wayleaves if you are to answer the question which we discuss in Chapters 6 and 7, namely is this an express or an implied wayleave and if so on what terms? This will be seen to be a crucial question.

5.13 Where licence-holders require voluntary permission

In certain circumstances it is not possible for a licence-holder to apply to the Secretary of State for a necessary wayleave. These restrictions are discussed further in Chapters 8 and 9 but in the context of the issues considered in this chapter it is worth noting that it can be expedient for the licence-holder to obtain permissions by agreement for electric lines across land which address requirements which a necessary wayleave could not meet. These restrictions largely relate to relevant dwellings. While the licence-holder always has the option to use its compulsory purchase powers, (discussed in Chapter 11), this is likely to be considered a sledgehammer to crack a nut where a low voltage overhead line is required to be installed and permission by voluntary agreement would be more timely and potentially more cost effective than following the compulsory purchase route.

5.14 Benefits to landowners granting easements

The granting of an easement, for a term of years or in perpetuity, for an electric line allows the full impact of the line and any injurious affection to be taken into account. If the development value is reduced by the presence of the line then that loss in value can be assessed under the compensation code[26] with a lump sum payment agreed for the loss in perpetuity. Where terms have been agreed for an easement but the consideration remains outstanding then it is possible for the landowner to agree with the licence-holder to make a joint reference to the Lands Tribunal to determine the amount of compensation. There would need to be a provision in the agreement as to the basis of compensation to be applied and normally this would mirror the terms of schedule 3 to the Act as if a compulsory purchase order with the equivalent rights had been acquired.

Where a term easement has been agreed this would be for the term of years specified. Allowance can also be made for any hope value at the time as at the valuation date the optimum use of the land may remain to be determined. Therefore in cases of lines across agricultural land it may be possible to agree the loss in agricultural use for the term granted of both the towers and the line and an allowance made for any planning permission likely to be granted. The extent of hope value will depend upon the particular circumstances and will clearly vary from land allocated as green belt to accommodation land on the edge of an urban area. Each case would need to be considered on its own merits. The lump sum payment would however properly compensate a landowner for the inability to use the land fully together with any injurious affection to the remaining land with reimbursement of all associated professional fees. The aim is that the landowner should be in no worse a position financially than if the line were not in place. With the grant of a voluntary wayleave there is the ability to terminate the wayleave, usually upon six or 12 months notice, and so the amount of compensation is limited to that duration.

---

[26] Section 5 of the Land Compensation Act 1961.

5.15 Benefits to licence-holder of easements

The fact that an express non-statutory wayleave may be terminated by notice makes it uncertain whether the line will remain across the land. There is therefore always the potential for the licence-holder to have to justify its position at a necessary wayleave hearing with the commitment of the resources which that implies. Of greater financial effect would be the situation where the Secretary of State did not grant a necessary wayleave and required the licence-holder to remove the line within a stated period of time. This would clearly be at the full cost of the licence-holder and, as the landowner would have terminated rights for the existing line it can be assumed that he would not agree to the line being placed in a similar position by the same means. Therefore a more circuitous route may be required. However, even if the landowner agrees to an alternative route that would still incur potentially significant costs for the licence-holder and therefore the security afforded by an easement would be an asset to that licence-holder's operations. Additionally where lines are for third party use, such as fibre optics cables, then it is possible to include terms for these cables within an easement by agreement. Accordingly, the taking of an easement should always be considered when a new line is in question.

Notwithstanding the ability to seek a consent under section 37 of the Act, there is always the potential of a hearing into the need for the line if any non-statutory wayleaves have been terminated for particular sections of it. When a multi-million pound investment on a particular line is in question there is clearly a commercial benefit in having security of tenure for the line before committing significant expenditure to it. An easement provides that certainty.

The same is also true where substantial works are planned for an existing line, even within the terms of existing easements and wayleaves.

5.16 Ancillary rights: relevance of ancillary rights

As we have seen, the terms of a wayleave will usually be supplemented by rights of access to enable the licence-holder to inspect and maintain the electric line. These rights are often called ancillary rights and that is the description we shall use for the remainder of this chapter. Such rights are needed if the basic permission granted by the wayleave is to be exercised efficiently by the licence-holder.

5.17 How are ancillary rights granted

In the case of necessary wayleaves, the court has decided that ancillary rights must always be implied into the wayleave.[27]

In the case of express non-statutory wayleaves,[28] ancillary rights will only exist if they are given by the terms of the documents signed by the current landowner/or occupier as grantor and the current licence-holder. If they are given by that document, they will usually be the subject-matter of an express provision. Occasionally, however, they may be implied into a voluntary wayleave on the basis of the apparent intention of the parties judged by the wording of the voluntary wayleave. Such an implication can be justified if it is necessary in order to give the wayleave business efficacy.[29]

In the case of an implied non-statutory wayleave,[30] ancillary rights will only be granted if the circumstances from which a wayleave is presumed to have arisen, likewise justify the deduction that such rights have also been granted.

5.18 Tree cutting

Where a wayleave exists for an electric line that is affected by trees in proximity to the line obstructing or interfering with the working of the line or constituting an unacceptable source of danger then the licence-holder would require permission from the owner of the tree to fell or lop the tree or, in the case of underground cables, cut back the roots. The statutory process for this is covered in Chapter 9 and it is invoked where agreement cannot be reached voluntarily. In those situations permission of the tree owner should be obtained by the licence-holder both on the extent and practice of tree cutting together with the loss in value of the tree. The owner of the tree may not have a wayleave with

---

27 *National Grid Co v Craven* see Chapter 8, section 8.4 for a fuller discussion.
28 See Chapter 6 for an explanation of this term.
29 The wayleave, as a contractual licence, is subject to the same rules as any other contract. For a fuller discussion see *Chitty on Contracts* 29th ed, vol 1, chapter 13.
30 See Chapter 7 for an explanation of this term.

the licence-holder if the tree is on land adjacent to that crossed by the electric line, or clearly would not have where the tree is on the route of a proposed electric line. It is possible for the tree owner to elect to carry out the tree cutting operations himself or his own appointed contractors, given suitable liaison with the licence-holder where this concerns an existing line, and recover relevant costs from the licence-holder. Certainly in the case of a proposed line across a country estate, where tree cutting operations are likely to exist already, it would be in the interest of the owner to arrange for his own staff or contractors to carry out the work at the most appropriate time to minimise the impact on the tree, where boughs are to be lopped or roots cut, and to reduce any damage to the land due to adverse weather conditions. Where agreement cannot be reached then it is open for the licence-holder to serve a notice on the tree owner to cut the trees. This is covered in more detail in Chapter 9.[31]

5.19 Surveys for exploration

Where it is necessary to carry out work to investigate the suitability of any land for the purposes of a proposed electric line, by commissioning a survey or undertaking invasive ground explorations, a licence-holder would require the permission of the owner and occupier of the land. Where this permission is not forthcoming then the licence-holder could rely on the rights provided for in paragraph 10 of schedule 4 of the Act requiring them to serve 14 days' notice of the intended entry and for the surveyor concerned to carry written evidence of their authority from the licence-holder. These powers are discussed further in Chapter 9.[32]

However where a survey is required to be carried out on land covered by a building, "or will be so covered on the assumption that any planning permission which is in force is acted on",[33] then a voluntary permission is required as it is not possible for a licence-holder to carry out a survey using their statutory powers.

The definition of building, where these powers could not be exercised, includes:

---

31 See Chapter 9, section 9.14.
32 See Chapter 9, sections 9.15 to 9.17 inclusive.
33 Paragraph 10 (3) of schedule 4 of the Act.

any garden, yard, outhouses and appurtenances belonging to or usually enjoyed with a building.[34]

There is no definition of appurtenances within the Act. In those situations the DTI applies the tests which are used to determine the definition of a dwelling[35] in order to assist the inspector to reach a reasoned conclusion on the definition of appurtenance. These are known as the use and comparability tests and in respect of a garden to a private dwelling require an examination to determine whether the land is used or was last used as a garden attached to a private dwelling or whether for wider commercial purposes. This would satisfy the use test. A test of comparability is to determine whether there is something special about the nature and situation of the land that distinguishes it from land in general. The nature of the ground would also be assessed to determine the principle characteristics of any structures, fixtures and fittings. The definition of building can therefore be seen to embrace a wider context than a dwelling.

It is possible for a garden to be physically separate from a house but there would need to be some proximity for it to fall within the definition of dwelling in paragraph 6 (8) of schedule 4 of the Act. Uses of land between the residential house and the garden would need to be determined and it is considered by the Department that even if a road separated a house from the garden this may still fall to be considered as being attached to the dwelling. However, a garden some distance away with conflicting uses of land between that area and the house, perhaps including field boundaries, roads and watercourses may determine that that garden is not attached to the house so as to fall within the definition of dwelling. Each case would therefore need to be considered on its merits and the inspector would listen to reasoned arguments as to whether any land physically separated but enjoyed with a house fell within the definition of dwelling for these purposes.

---

34 Paragraph 10(6) of schedule 4 of the Act.
35 See paragraph 6(8) of schedule 4 of the Act in relation to wayleaves for proposed lines.

Non-Statutory Wayleaves: Voluntary Wayleaves

6.1 Outline of chapter

This chapter explains what a voluntary wayleave is and the various forms a voluntary wayleave can take. The discussion will enable the reader to understand first the distinction between a voluntary wayleave and an implied wayleave. These together form the category of non-statutory wayleave. This category of wayleave must in turn be contrasted with a necessary wayleave properly so called. In this chapter and in Chapter 7 we consider the crucial importance of correctly identifying the type of wayleave in existence in any given situation before implementing procedures considered in Chapters 9 and 10.

6.2 Technical meaning of "necessary"

In practical terms all wayleaves are necessary. Without any wayleave, a licence-holder cannot install an electric line on the land if the land belongs to someone other than the licence-holder. As discussed in Chapter 5,[1] without a wayleave the installation of an electric line on someone else's land is a trespass.

In legal terms "necessary" has a very specific meaning. It means a wayleave granted by the Secretary of State in the exercise of the powers granted by schedule 4 to the Act. The statutory definition is considered in Chapter 8. The statutory powers are the subject-matter

---

1 See Chapter 5, section 5.3.

of Chapters 9 and 10. Insofar as a necessary wayleave has been granted by the Secretary of State, notwithstanding the absence of any consent on the part of the landowner or occupier, it can be said to have been granted under compulsion. Indeed the rules governing this process, as we shall see, refer to "compulsory wayleaves". Compulsory and necessary are one and the same.

6.3 Technical meaning of "voluntary"

By contrast, a voluntary wayleave is one made by agreement between a landowner or occupier and a licence-holder. That follows the commonly understood meaning of voluntary, namely a willing and intended act. As we have seen the legal category into which a wayleave falls is a contractual licence. In the next chapter we consider wayleaves which arise by implication. An implied wayleave is also a contractual licence. It would therefore be technically correct to describe an implied wayleave as a voluntary wayleave. However, that would not accord with the perceived understanding of the electricity industry. We will therefore confine the term voluntary wayleave to those wayleaves which arise because the parties have consciously come to terms. Wayleaves that arise by implication are discussed in the next chapter.

One other point needs to be made at this preliminary stage; a wayleave can be voluntary even though the parties have not necessarily hammered out each individual provision by specific negotiation. If standardised terms are adopted by mutual consent then the wayleave is still voluntary. The same is also true if the wayleave refers to terms (particularly as regards payment) which are negotiated by representative national bodies.[2] To have a clearer understanding of the distinction between the wayleaves discussed in this chapter and the next, it would be better to talk about wayleaves which are either express or implied.

6.4 Types of express wayleave

In considering express wayleaves, we need to subdivide the category as follows:

---

2   See the examples mentioned in Chapter 14, section 14.2.

- express oral wayleaves and
- express written wayleaves.

The person who grants an express wayleave is the grantor. The person to whom the wayleave is granted is the grantee.

The grantee will always be a licence-holder. The terms express and grantor require further comment.

6.5 Technical meaning of the word "express"[3]

In legal terminology express means spelt out in full or explicit. Expressly agreed means the parties have contracted on terms they have actually negotiated and settled. This will include any terms incorporated by reference and adopted as part of the "express agreement". So a wayleave granted because the parties have actually turned their mind to it is express in that sense. However, express can also refer to the terms of the wayleave. In Chapter 5[4] we considered the basic legal hallmarks of a wayleave. Over and above the bare essentials there may be ancillary rights such as rights of access for inspection and maintenance. Again, these may have been expressly agreed. Even in an express wayleave, however, a term may be implied. Although the parties have clearly decided to contract and an express wayleave has obviously been granted, the fine details may not have been fully worked out. If from the apparent intentions of the parties judged by such terms as have been expressly agreed it is clear that something else was intended, but was not totally spelt out in the wording of the wayleave, the Court will imply a term to give business efficacy to the wayleave.[5] This process of implication is similar to, but nevertheless distinct from, the process of reasoning which leads to the conclusion that the whole wayleave has arisen by implication. That we discuss in the next chapter.

Nevertheless, there are limits on the court's ability to supplement the language of a contract in order to make it work commercially. A

---

3   In contrast to the vernacular meaning of speedy or accelerated as in "express planning consent".
4   See Chapter 5, sections 5. 8 and 5.9.
5   A full discussion of this subject is outside the scope of this book. The reader is referred to *Chitty on Contracts* 29th ed, Chapter 13.

term[6] ought not to be implied unless it is in all the circumstances equitable and reasonable. But this does not mean that a term will be implied merely because in all the circumstances it would be reasonable to do so or because it would improve the contract or make its carrying out more convenient. The touchstone is always necessity and not merely reasonableness. The term to be implied must also be capable of being formulated with sufficient clarity and precision.

This is an important limitation to bear in mind when considering whether the express wayleave contains a term authorising a review of the wayleave payment and, if so, on what terms. There is no automatic right to a review. Unless provision for a review has been expressly agreed, there is no right to one. It is of course open to the parties to agree to a review nevertheless. However, in those circumstances, a new contract will have been created. If no new document is signed, the legal analysis would be that an express oral wayleave had been granted on the terms of the provisions of the previous express wayleave except as to the new payment terms.

6.6 Express wayleaves: who can grant them?

Before considering the various sub-categories of express wayleave in more detail, we need to explore the fact that we have in this chapter and elsewhere in the book used interchangeably landowner and/or land occupier to mean the persons who can actually grant a wayleave.

The freehold owner of any land[7] can grant any permission to use all or part of his land as the freehold owner sees fit. A licence-holder may wish to be assured that a person claiming to be the freeholder owner does in fact have that quality of legal estate. This can easily be demonstrated by reference to the Land Register if the title is registered, or alternatively if the title is unregistered by the production of the conveyance by which ownership was passed to the grantor. If the conveyance is less than 15 years prior to the date of the transfer then any preceding conveyance will also need to be produced until a good root of title can be shown.[8]

---

6 The text is a quotation from Paragraph 13–009 of *Chitty on Contracts* mentioned above and the authorities cited in that paragraph.
7 For an explanation of freehold please refer to Chapter 5, section 5.7.
8 For a full discussion the reader is referred to *A Practical Approach to Conveyancing*, 6th ed, Robert Abbey & Mark Richards.

A tenant has the right to the immediate possession of the land which is the subject-matter of the lease or tenancy. Because a wayleave is not the grant of a right or interest in land[9] but rather a mere permission, a tenant in principle has the right to grant such a permission. The general rule is that a person can grant any lesser privilege or interest over land that can be properly derived from the one that he himself enjoys. A tenant carves out of the freehold interest a specific period (the term) during which the tenant enjoys the possession of the land. A convenient allegory for estates and interests in land is that of a food chain. The freehold owner is the animal at the top of the food chain. A visitor who comes onto the land with permission for a limited period and then leaves would be at the very bottom. In between there can be various grades of tenant depending upon the length of the term they enjoy. The term is the right to enjoy the demised premises and possess it to the exclusion of the person immediately above you in the food chain. A wayleave being a mere permission is of a lower grade than a lease or a tenancy and therefore a lessee or a tenant can in principle grant one subject to any actual provisions to the contrary in the lease or tenancy agreement.

It is for that reason that the Act recognises that the persons with whom you have to deal in relation to wayleaves can be both a landowner and a land occupier. However, if an express wayleave is being taken from a tenant, then the licence-holder must exercise some care to be certain that the tenant not only has the right to grant the wayleave, but also the power to grant it. Many leases or tenancies restrict the ability of the tenant to grant further rights over the demised premises. The reasoning behind this is that the freehold owner does not want eventually to have the land back with new rights granted over it.

6.7 Express wayleaves: oral-specific points

It is our suggestion that the burden of proof required on any person asserting that the original basis on which permission to install an electric line was granted was an express oral wayleave (that is to say an express wayleave not created by a document) would be very high indeed and practically incapable of being discharged.

Given the industrial background to the modern development of wayleaves, a document is likely to be the basis of the original rights.

---

9   See the discussion in Chapter 5, sections 5.10 and 5.11.

Apart from any other factor, the licence-holder will need clarity on what other rights it has over and above the basic permission to keep infrastructure on land. Rights of access maintenance and inspection are the usual candidates for ancillary rights. Hence it is highly improbable that the licence-holder would have been prepared to acquiesce in an express wayleave granted orally as an initial method of authorising an electric line on land.

However, as we have discussed above, an express oral wayleave may arise in circumstances where the parties have begun their dealings with an express written wayleave but have continued it less formally. In those circumstances an express oral wayleave may well arise.

There are dangers in such informality, particularly when it comes to the procedures discussed in Chapters 9 and 10. Our recommendation would always be that all express wayleaves should be in writing.

6.8 Express wayleaves: written-specific points

The position is clear if, as between the current landowner and the current licence-holder, there is a wayleave by an agreement in writing to which they are the actual signatories as the rights enjoyed by that licence-holder will be set out in the agreement. An owner or occupier of land's right to terminate (if any) will likewise be found in the agreement. If there is no right to terminate the wayleave on notice, then the wayleave will usually be for a fixed term. Express written wayleaves would usually incorporate the national payments scale discussed in Chapter 14. That fact does not make the wayleave any less voluntary. It is not necessary that every single term be negotiated afresh as if it were a provision in the first wayleave ever granted.

A wayleave in perpetuity with no right to terminate is not something the authors have personally encountered and, while it may in theory be possible to have such a permission, it is unlikely to arise in practice. However permanent as against the grantor, the wayleave would not survive a change of ownership or occupation of the land. Moreover, it would be arguable that such a wayleave had moved beyond the realms of privilege and permission and had in fact become an easement, which is a very different category.[10]

---

10 See Chapter 5 section 5.10.

One other term that may be found in an express written wayleave is a provision to the effect that if at any time a landowner or occupier considers that there has been, or can demonstrate that there has been a diminution in the capital value of the land as a result of the existence of the wayleave, he can then apply to the Lands Tribunal for compensation.[11] On payment of such compensation, as the Lands Tribunal may award, the wayleave may also provide that the landowner or occupier has to grant an easement to the licence-holder in substitution for the wayleave.[12] A crucial point to bear in mind is that an occupier (as opposed to the freeholder) may be subject to a restriction preventing this.

A variation on this theme is that the landowner becomes entitled to compensation if he obtains a planning consent for development which cannot be implemented because of the existence of the wayleave.

6.9 Express wayleaves: statutory substitution of licence-holder

Finally, we comment upon the effect of the changes brought about by the Utilities Act 2000.[13] The fact that with effect from 1 October 2001 there were changes in asset ownership in consequence of the reorganisation of the licensing regime does not bring a voluntary wayleave to an end. It merely substitutes one licence-holder for another. Everything else remains the same. It is not necessary, therefore, to consider in those circumstances the extent to which the general contract law of assignment and novation is applicable to express wayleaves. The statutory provision exempts an express wayleave from the common law rules insofar as they might otherwise apply in such cases.

---

11 See Chapter 16.
12 See Chapter 5 generally and specifically Chapter 5, sections 5.10 and 5. 11 for the difference between an easement and a wayleave.
13 See Chapter 4, section 4.4.

Non-Statutory Wayleaves: Implied Wayleaves

7.1 Outline of chapter

An implied wayleave is a wayleave which is deemed to have arisen by implication from circumstances and the conduct of the parties, whether the parties specifically intended it to arise or not. The parties will not have expressly negotiated terms. Nevertheless they will be taken as having agreed to a new contractual licence. This chapter considers how you can identify if an implied wayleave has come into existence.

7.2 Circumstances from which wayleaves may be implied

The usual circumstances in which an implied wayleave will arise are where a wayleave has originally been granted expressly in writing but a new landowner now owns the land over which the wayleave runs. The wayleave automatically lapses on the change of owner. However, a new contractual licence can be implied from the course of dealings. In the absence of any evidence to the contrary, the terms of the original wayleave are assumed to continue as between the new owner and the licence-holder.

Similarly an implied wayleave can arise if, notwithstanding the fact that an electric line has been installed, there is no apparent applicable agreement in writing giving the necessary permission and no evidence of an express oral wayleave. These circumstances can arise when a fixed term express wayleave automatically expires on the completion of the fixed term but the electric line is not then removed. A fresh contractual licence may nevertheless be implied from the conduct of the parties.

The crucial enquiry to make is whether there has been a payment of any relevant fee notwithstanding any change of ownership, or automatic expiration or lapsation. If a fee is paid in consideration for a permission to lay or install an electric line on land, you have sufficient evidence for an implied wayleave.

What then has to be found is any documentary evidence such as previous wayleaves, invoices or letters which show what the parties intended by way of detailed terms.

7.3 Other terms

A bare permission may not be sufficient for the purposes of the parties and other terms will be needed. To determine what those other terms may be, the first task, therefore, is to find out whether any kind of document has ever been applicable in the past. With the nationalisation of the electricity industry in 1947, the Central Electricity Board and subsequently the Central Electricity Generating Board took grants of wayleaves from thousands of landowners in a common documentary form. If the evidence that proves the continuation of the payment also cross refers to such standard terms, then this is the evidence needed to prove that the terms in the original document have been carried over into the new contractual licence.

Unless any information can be found which shows that the terms of the original agreement were in fact varied by agreement, those will be the terms which the parties will, in all likelihood, have carried over into their more informal arrangement.

However, if there never was an original document then the terms of the implied wayleave will be confined to what can be deduced from the correspondence or invoices created in the circumstances from which the wayleave is implied.

7.4 Identification of licence-holder

As has been mentioned, as a result of the Act and the Utilities Act 2000, the principle of automatic expiration or lapsation does not apply to changes on the licence-holder's side. The statutory successors to the original grantee inherit the benefit of the wayleave. However, remember that a licence-holder is a broad definition and the actual licence-holder may not be the company one would immediately recognise as the electricity provider. To summarise the discussion in sections 2.2 and 2.3, between 1947 and 1990, a wayleave would be granted to an Area Electricity Board which was the statutory predecessor to the post-nationalisation licence-holders. The Area Electricity Boards operated within defined regions. From 1990 their geographical boundaries became the franchise areas for the newly created Distribution Network Operators. The first enquiry is, therefore, to which of these geographic areas does the wayleave relate? The areas are illustrated in Appendix 2. In England and Wales there are 12 such areas. Full corporate names may be obtained from the Ofgem website.

At the boundaries of these areas, more detailed investigation is needed. One of the ways of establishing which licence-holder owns a particular overhead line can be assessed by the type of overhead line support[1] and, in some cases, identification labels. All electricity equipment will have danger of death notices attached and, in addition, identification labels providing the name of the circuit and support number. With lines operating below 132,000 volts, which are the majority, there is likely to be only the tower or pole identification number and the danger of death notice. Any enquiry for lines on the regional boundaries needs to be made with one or other of the relevant licence-holders as to whether their lines cover the land in question. In the minority of cases where lines may cross the geographic boundary, the line support number should be provided when making an enquiry of the distribution network operator for confirmation as to which company is responsible for maintaining and operating the lines.

Enquires should be made to the correct licence-holder as there is no obligation upon one licence-holder to pass enquiries to another. The importance of making correct enquiries becomes clear when the procedures discussed in Chapters 9 and 10 are considered.

When looking for details of which licence-holder has been paying the wayleave payment, the crucial date from which you need to trace a payment history is 1 October 2001.[2]

---

1 See Chapter 3, section 3.3 for a definition of this term.
2 As discussed in Chapter 4, section 4.4.

Part 3
Public Control

The Statutory Menu Relating to Wayleaves, Consents and Easements

8.1 Outline of chapter

This chapter explains:

- when the opportunity to obtain a necessary wayleave arises
- what the necessary element means in the statutory phrase "necessary wayleave"
- the key point to bear in mind about necessary wayleaves
- how a necessary wayleave compares and contrasts with a non-statutory wayleave
- the interaction between wayleaves, consents and easements.

8.2 When the opportunity to obtain a necessary wayleave arises

Although the legal unit of currency within the electricity industry is a contractual licence, which is by definition terminable, this does not give landowners the right to hold licence-holders to ransom. If a voluntary or an implied wayleave ends for any reason, the landowner may give the licence-holder a notice to remove the electric line. The licence-holder then has three months in which to apply for a necessary wayleave. The notice to remove is not effective unless the relevant

65

procedures have been completed. Equally, if there never has been a voluntary wayleave and it is "necessary and expedient" that there should be one, the Act provides a mechanism, which the licence-holder can initiate, for obtaining a necessary wayleave. These procedures are discussed in more detail in the following chapters.

The point about a necessary wayleave is that it is granted by the Secretary of State on the application of the licence-holder[1] and will be granted on terms including a term relating to the period for which it lasts. It then has all the features mentioned below.

It can be granted even if an owner or occupier of land objects.

8.3 Statutory definition

The Act defines[2] a necessary wayleave as a "consent" given by an owner of land to a licence-holder "for the licence-holder to install and keep installed an electric line on or under the land and to have access to the land for the purpose of inspecting, maintaining, adjusting, repairing, altering and replacing or removing the electric line".

Note that the statutory definition does not set out all the characteristics of a wayleave for general purposes. An understanding of what a wayleave is and how voluntary and implied wayleaves arise[3] is taken as read.

To understand fully the statutory definition we must review the decision in *British Waterways Board* v *London Power Networks plc* [2003] 1 All ER 187.[4] In that case, the Court has considered the definition of land for the purposes of the Act. The issue was whether the Secretary of State had the power to grant a necessary wayleave permitting the installation and maintenance of electricity cables through service tunnels running under Mill Wall Cutting, Marsh Wall, London E14. The Act has no specific definition of its own and therefore the definition given by the Interpretation Act 1978 applies.[5] It was not

---

1 Schedule 4, paragraph 6(3).
2 Schedule 4, paragraphs 6(1) and 6(2).
3 See Chapters 5, 6 and 7.
4 We also consider the case in connection with a different aspect of the decision in Chapter 9, section 9.12.
5 See Chapter 5, section 5.6.

disputed that service tunnels themselves were to be treated as land.[6] It was therefore held that an electric line in a service tunnel was an electric line "on under or over land" for the purposes of the Interpretation Act 1978. Accordingly since the land was not one of the categories over which the Act specifically states that the Secretary of State has no jurisdiction to grant a necessary wayleave, he was empowered to grant one in that case.

8.4 The legal characteristics of necessary wayleaves

A necessary wayleave is always granted in writing. Accordingly, there is no scope for an implied necessary wayleave to arise, whereas many if not most non-statutory wayleaves are implied.

A necessary wayleave passes through successive ownerships. The Act achieves this effect by stating[7] that any necessary wayleave granted under the Act shall bind any person who is at any time the owner or occupier of the land.

So far as rights ancillary to the essential permission are concerned in an unreported case decided by the Court of Appeal *National Grid Co plc v Craven* it was decided that "to install" for the purposes of a necessary wayleave includes "to enter and construct with a view to installing" and so the necessary wayleave incorporates automatically all requisite rights of access.[8] Such rights are not automatically included in a voluntary wayleave. If the express terms do not give these rights then you would have to argue that such rights ought to be implied having regard to the usual general principles of contract law.[9]

---

6   The parties regarded the point as settled by *Elitestone Ltd v Morris* [1997] 2 All ER 513 in which the House of Lords reviewed the law relating to chattels and fixtures brought onto land and the point at which fixtures become part of the land itself. The House of Lords used their judgments in this case as a vehicle for restating the fundamental principles.
7   Schedule 4, paragraph 6(6).
8   *National Grid Co plc v Craven*: I am indebted to an article dated 26 November 2003 entitled "Enforcement of wayleaves — Getting it Right!" by David J. Williams for bringing this decision to public attention.
9   So far as the rules on implied terms relate to wayleaves see Chapter 6, section 6.5 and Chapter 7, sections 7.2 and 7.3. For a fuller discussion of the topic more generally see *Chitty on Contracts* 29th ed.

Finally, the licence-holder who has the benefit of a necessary wayleave is deemed to be a landowner for the purposes of the Mines (Working Facilities and Support) Act 1966. That Act deals with the granting of rights for mining operations and the effect of this provision is that the licence-holder is an interested party for the purposes of that Act.

8.5 Registration of necessary wayleaves

A necessary wayleave "shall not be subject to the provisions of any enactment requiring the registration of interests in charges over or other obligations affecting land."[10]

This means that you cannot identify from searches at the Land Registry or a local land charges search from a local authority or from a search at the Land Charges Department whether or not a necessary wayleave has been granted. However, if a landowner acquires land in respect of which a necessary wayleave has been granted and that wayleave is current at the date of acquisition, an owner or occupier of land will acquire the land subject to, and be obliged to accept the wayleave. Therefore, to prevent difficulties, a landowner must make specific enquiries of a vendor before acquisition and notify the licence-holder of the change of ownership afterwards.

8.6 Temporary continuation of wayleaves by statute

Finally we need to consider the nature of the wayleave implied by paragraph 8 of schedule 4 to the Act. As explained above, the mere termination of a wayleave does not automatically require a licence-holder to remove the electric line from the land. There are further processes which have to be completed before an obligation to remove arises. During that period, paragraph 8 of schedule 4 to the Act provides that notwithstanding the termination of the express or implied voluntary wayleave by notice or expiration of the term (or in express cases only change of ownership) a wayleave will be deemed to continue on the same terms and conditions as previously unless and until the process as described in Chapters 9 and 10 have been completed.

---

10   Paragraph 6(6)(a) of schedule 4 to the Act.

The question then is what, if any, rights arise if the landowner/occupier demands and/or accepts further wayleave fees during this temporary confirmation period? Second, to what extent, if at all, are any terms applicable to the wayleave, either expressly or by implication, still relevant to the relationship between landowner and/or occupier and licence-holder? This will be important from the point of view of identifying the extent to which the landowner/occupier, for instance, must continue to grant rights of access to the licence-holder.

If wayleave payments continue after service of a notice to terminate or after automatic termination and are accepted by an owner or occupier of land before a notice to remove, then we suggest that a new implied wayleave arises.

The terms of that new wayleave would need to be identified by reference to the actual factual context in which the payments were made. The only way to prevent such an implication arising would be to make it clear that all an owner or occupier of land and the licence-holder are doing is treating the now terminated wayleave as effectively continuing to regulate their dealings pending the service of a notice to remove and the completion of the statutory process. Strictly, the statutory continuation period does not begin until three months after the notice to remove is served so there could be a period between termination or lapsing or expiration of the wayleave and expiration of the notice to remove when the legal position is not clear.

Best practice would be that the parties expressly acknowledge and notify each other of their intentions by correspondence.

If wayleave payments continue during the implementation and completion of the procedures discussed in Chapters 9 and 10, the question is, what is the legal basis for the continuation of a wayleave payment? By definition there is no longer a non-statutory wayleave. Is there, then, a trespass? It would seem odd to suggest so because the Act itself presupposes the continuation of the electric line. Where is the "unjustifiable intrusion"[11] onto the former grantor's land? This is a matter we consider further in Chapter 22. However, on the assumption that the financial remedies for trespass[12] are not available then this must be a matter for consideration when the Lands Tribunal awards compensation.

---

11 See Chapter 5, section 5.3.
12 See Part 5 of this book.

8.7 Link between wayleaves, easements and consents

We refer in detail to the requirements of non-statutory wayleaves at Chapters 6 and 7, statutory wayleaves in Chapter 10, rights granted under easements at Chapter 15 and planning and ministerial consents at Chapter 12. There is therefore a need to understand how the matrix of these work and how ancillary rights, including access to survey and for tree cutting near lines operates in practice. Tree cutting and surveying we will consider following an overview of the interaction between wayleaves, easements and consents.

Where the necessary wayleave process is denied to the licence-holder, for example if a new overhead electric line were to cross land covered by a dwelling, as discussed in Chapter 9, then the options open to the licence-holder are to negotiate terms for an easement, as discussed in Chapter 15, by agreement with the owner. This would allow extended terms to be considered beyond that available under a wayleave. Alternatively the licence-holder could pursue a compulsory purchase order under schedule 3 of the Act as discussed in Chapter 15. However in both cases it is not possible to obtain the minister's consent for a proposed line, and thereby express planning consent, as discussed in Chapter 12, until all permissions required from owners and occupiers that will be affected by the proposed electric line have been obtained. A situation could also arise where a proposed overhead electric line could impact on trees on growing on land adjacent to the land to be crossed by the proposed line or where existing trees could interfere with the construction and subsequent operation of the line.

A licence-holder has the obligation to maintain an efficient, economic and co-ordinated system under either a transmission or distribution licence and have regard to their environmental obligations including mitigating the effect of their proposals. In acquiring rights from affected owners and occupiers, or in the absence of voluntary agreement, from the Secretary of State, they also have an obligation to pay fair compensation to the owner/occupier. The process the licence-holder must follow is to ensure that the correct owners and occupier are identified, the status of the land, in relation to applications for necessary wayleaves, is verified and the extent of their proposals fully examined. This is to take account of effect on all affected owners and occupiers as well as mitigating the licence-holder's proposal on the environment.

In the case of an existing line it is essential for an owner/occupier to verify if the rights of the licence-holder comply with the terms of any agreement they have, whether those terms are binding on that owner/occupier, whether the rights required comply with the terms of the planning and minister's consent for the line and whether the situation in the locality has changed from when this consent was granted such that a wider consideration be taken account of, with respect to the minister's consent, which is discussed further in Chapter 12.

The interplay between the various rights and their requirements requires full consideration before determining what approach to take in any given circumstances.

8.8 Tree lopping and felling

Rights under both statutory and non-statutory wayleaves include rights for access to and maintenance of an electric line. However where the line is affected by trees growing in the vicinity of the electric line then a notice needs to be served on the owner of the trees as discussed in Chapter 5. A wayleave would not provide rights to cut trees, although an easement may, and the initial requirement is for the licence-holder to require the tree owner to cut them. Where the tree owner objects they have the opportunity of being heard by the Secretary of State or they could terminate a voluntary wayleave and then serve a notice requiring the removal of the electric line to avoid the interaction between the electric line and the trees. The statutory provisions in relation to tree cutting and termination of wayleaves are covered in detail in Chapter 9.

8.9 Environmental obligations

At any stage the environmental obligations[13] of the licence-holder need to be taken into account. A licence-holder is required to take into account the effect of any of their proposals on flora, fauna and features of historic, architectural or archaeological interest and to mitigate the effect of their proposals on such features. This goes beyond the

---

13   Schedule 9 to the Act.

contractual relationship with a landowner and requires a wider remit to be considered even where rights exist for an electric line under a wayleave, easement or minister's consent.

8.10 Conclusion

Rights for an electric line may be covered by a statutory or non-statutory wayleave or by an easement across private land. In addition a valid planning and minister's consent must exist for the specific electric line for it to be lawfully placed or retained. Where trees interfere with an electric line, or where a survey is required for a new line, then certain procedures must be followed and the opportunity provided for all aspects to be considered.

The opportunity exists for a review of existing rights for an electric line to be undertaken in a number of circumstances. Where there is a specific wayleave agreement under its terms or through termination of the agreement or by serving a notice to remove the line where the rights have not transferred to a new owner. Where the rights are under an easement a review could be commenced under a specific review clause and even where a party is not able to action a review as described, or does not have private land affected, they could still request a review of the minister's consent. All cases would need to be considered on their individual circumstances with the intention here being to give an overview of the interaction and matrix of options available for both licence-holders and landowners.

Applications to the Secretary of State for Statutory Rights

9.1 Outline of chapter

This chapter explains in detail the procedures embedded in schedule 4 to the Act.

In relation to wayleaves this chapter falls into four parts. Sections 9.2 to 9.6 inclusive form Part 1 which considers how in relation to an existing electric line, a landowner must begin the process which will lead:

- either to the termination of a wayleave and the removal of an electric line or
- to the grant of a necessary wayleave or
- to the replacement of an existing necessary wayleave by another.

Section 9.7 forms Part 2, which considers how a licence-holder begins the process for obtaining a necessary wayleave on its own initiative. This process applies only to new electric lines. What is common to both the situation reviewed in Part 1 and that discussed in Part 2 is that it is always the licence-holder who makes the application to the Secretary of State.

Sections 9.8 to 9.13 inclusive form Part 3 which describes how such an application is commenced.

Finally sections 9.14 to 9.17 inclusive form Part 4 in which we discuss the rights a licence-holder can employ or apply for in relation to the cutting of trees and surveying private land.

9.2 DTI guidance

The Secretary of State's jurisdiction has been remitted to the Secretary of State for Trade and Industry and so is exercised via the Department of Trade and Industry. The procedures relating to necessary wayleaves embedded in schedule 4 to the Act have been supplemented by guidance issued by the Department of Trade and Industry. A flowchart published by the DTI appears at Appendix 11A. We have imposed a six stage structure upon it. To relate the three-part structure of and the scope of this chapter and the scope of the following chapters to the flow chart:

- this chapter considers stages 1–4 inclusive. Stages 5 and 6 (the actual Wayleaves Hearing and its aftermath) are considered in the next chapter
- of the four starting points in Stage 1, the three beginning "existing line" are considered in Part 1 of this chapter
- in relation to a new line, the sequence involving the service of notice by the licence-holder is considered in Part 2 of this chapter
- the sequence involving compulsory purchase is reviewed in Chapter 11.

Citations of numbered paragraphs in the DTI guidance will follow the convention DTI [No].

Paragraph 8 of schedule 4 to the Act contemplates the following distinct situations in which the need to make a necessary wayleave application might arise, namely:

- a voluntary wayleave expires because the fixed term for which it was granted has come to an end (paragraph 8(1)(a))
- a wayleave whether voluntary or implied or even a necessary one is brought to an end by notice (paragraph 8(1)(b)
- a voluntary or an implied wayleave has automatically terminated on a change of ownership or the death of a grantor (paragraph 8(1)(c).

It is important to be clear about which is the relevant situation to you because on that question depends the answer to the questions: how many notices do you serve and when? The crucial point for present purposes is that if you want to have any prospect of causing an electric line to be removed, you must be able to show that you have validly brought to an end whatever form of wayleave is already in existence or that whatever form of wayleave was in existence has for the reasons previously discussed automatically lapsed or expired. In any event you will always need a notice to remove.

9.3 General points about notices to terminate

Wherever a wayleave is capable of being terminated by notice, a wayleave is terminated by the landowner or occupier serving notice of the correct length on the correct licence-holder. The date of termination is the date on which the notice period expires.

The correct length of notice will be established by the terms of the wayleave. For a voluntary wayleave this means consulting the document signed by the current landowner or occupier and the current licence-holder. For an implied wayleave this requires a consideration of the context in which the wayleave was implied to see what notice may be taken to have been agreed.

Provided a wayleave is not expressly made incapable of being terminated by notice, there will not necessarily be a problem if no notice period is specified. This point applies equally to express and to implied wayleaves. If no notice period is specified and the wayleave is periodical (*not* for a fixed period), a notice period will be implied which has to be reasonable. There is no definition of reasonable in the cases beyond "what is reasonable depends on the circumstances." DTI guidance suggests that six months is reasonable. That is the recommended best practice.

Pitfalls to avoid are:

- serving notice to terminate a wayleave on the wrong party because you have not correctly identified the licence-holder bearing in mind the changes in ownership which may have occurred[1]
- overlooking the fact that there may now be an implied rather than an express wayleave and so there may have been an alteration to the length of the wayleave by reference to custom and practice

---

1   See the discussion in Chapter 4 section 4.4, Chapter 6, section 6.9 and Chapter 7, section 7.4.

- there is no prescribed form for a notice to terminate. If the terms of a wayleave lay down a form then that must be followed otherwise any notice in writing clearly describing the land and the electric line and giving the right information as to termination will be sufficient. A notice in writing is necessary regardless of the category of non-statutory wayleave.[2]

9.4 General points about notices to remove

There is no prescribed form for a notice to remove. The DTI guidance suggest that such a notice merely has to refer to the parties and the land and the electric line correctly and be unambiguous about the requirement to remove the electric line by a due date.

9.5 One notice procedure

In any case where a wayleave has lapsed or expired or is due to expire, nothing is needed to bring it to an end. The landowner goes straight to a notice to remove. Timing, however, differs according to the situation.

If a fixed term is due to expire then a notice to remove may be served within three months of the due date or at any time afterwards (Paragraph 8(2)(a)). For the reasons explained in Chapter 6 this would only ever be applicable in the case of an express wayleave.

If the wayleave has lapsed on death or change of ownership then the notice to remove can be served at any time after that event has occurred.

9.6 Two notice procedure

In any other case the wayleave must be first terminated in accordance with its terms. Only after termination can a notice to remove be served. It will be seen therefore that only in the case of express fixed term wayleaves can a notice to remove be served so as to coincide with the date on which the wayleave expires by completion of the fixed term.

---

[2]   Cross refer to Chapters 6 and 7.

9.7 Initiative taken by licence-holder

Under paragraph 6(1) of schedule 4 to the Act, a licence-holder has the right to request from any landowner or occupier a wayleave sufficient for its purposes. If having been given at least 21 days the landowner or occupier has not done so voluntarily, the licence-holder may apply to the Secretary of State for a necessary wayleave.

By definition, therefore, this procedure will only be invoked in respect of new lines.

The pitfall for the licence-holder is that it serves on the wrong party. If the title to land is registered then the identity of registered freehold proprietor will be easy to discover. However, anyone with a lease or a tenancy has the right to immediate possession. Such a person would also have to be served. If the tenant's interest is not held through the medium of a registrable lease then the full range of relevant interests can only be identified by enquiries in the locality.

9.8 Stage 3

The procedures in stage 3 are applicable in all of the above circumstances. However, the requirements for what information has to be supplied with an application do differ. Where requirements for information are applicable to specific circumstances only, this will be made clear; otherwise what follows is relevant to all of the above circumstances equally.

There is no prescribed form for an application to the Secretary of State. A letter containing the requested information will suffice. Indeed, the DTI states that it will accept an application by email with any supporting documentation that cannot be transmitted electronically to follow by post as well as by fax. However, in both cases the DTI suggests that licence-holders also send a hard copy by post and check with the Department's Wayleaves Manager that the fax or email has been received. All necessary wayleave applications are stamped and dated on receipt.[3] Note that the date and stamping on receipt is crucial from the point of view of working out whether the application has been made on time.

---

3   DTI (3.12) and DTI (3.13).

The reference to the Department's Wayleaves Manager is reference to a civil servant within the Department of Trade and Industry who is in charge of managing all the administration relating to such application. The relevant contact details are in Appendix 28.

9.9 Contents of the application: general

What should the letter of application contain?[4] The contents are not prescribed by the rules or any other statutory instrument or the Act. However, the DTI does request that certain information be included within the application and we recommend that that request is met as a matter of best practice. The information is set out in Appendix 11B.

There are two comments we need to make on the requirements set out in Appendix 11B. The first is that licence-holders make the application for a necessary wayleave and so, the first statement which the DTI will receive as to whether or not there are negotiations and whether the application should be held in abeyance is a unilateral statement on the part of the licence-holder. The landowner/occupier will not have become officially involved at this stage. The second comment is in relation to the definition of electric line. The statutory definition of electric line[5] is set out in Chapter 4. The DTI states that it considers that an electric substation is "electrical plant" as defined separately in section 64. In consequence, it takes the view that Schedule 4 is not applicable to cases where the licence-holder wishes to install or construct electrical plant. That would be a matter for a consent under section 36 of the Act.

9.10 Additional requirements: existing lines

Where the application concerns an existing line, the further information listed in Appendix 11C is also requested.

---

4   DTI (3.17).
5   Section 64 of the Act.

9.11 Additional requirements: new lines

Where the application concerns a new necessary wayleave for a line not yet built (the landowner/occupier having failed to grant one on notice) then in place of the additional information set out in Appendix 11C, the information set out in Appendix 11D is requested by the Department of Trade and Industry.

The significance of the information in Appendix 11D is that the Secretary of State's ability to grant a necessary wayleave in the first place is restricted if there are dwellings in question. We consider the dwellings exception further in section 9.13.

9.12 Pitfalls to be avoided

The DTI guidance offers a number of recommendations and draws the attention of licence-holders in particular to a number of pitfalls which have to be avoided.

The first point is that the requested information is a suggested minimum best practice requirement. The DTI does say that the lists in its guidance are not exhaustive. All necessary wayleave applications are different and it is for the applicant to judge whether there is other information that might be relevant. Notice that the responsibility for making an application in the most appropriate form rests firmly with the licence-holder.

In cases where a previous necessary wayleave application has been made, eg where the previous application was made prematurely or there are other current applications relating to the same line the Department's reference number should also be included. The Department's reference number allocated to a particularly necessary wayleave application should also be included in all subsequent correspondence.

While not a requirement of the legislation, the Department recommends that the licence-holder copies its necessary wayleave application to the owner and/or occupier and explains why the application has been made. The DTI further recommends that as a matter of good practice, licence-holders and/or occupiers send their applications and notices by recorded or special delivery (as these methods require an acknowledgement of receipt). The DTI also suggests that licence-holders contact the DTI's Wayleaves Manager if confirmation of receipt has not been given prior to the end to the three month period for making necessary wayleave applications. Although the Department will endeavour to acknowledge new applications on receipt, there may be occasions where a response is delayed. There is no requirement for the owner and/or occupier to copy notices to the Department.

One further comment is necessary at this point: the recommendations of the DTI set out above presuppose that the licence-holder is effectively making the application for the necessary wayleave at the beginning of the three month period, rather than towards the end. It is only necessary that the application be made before the expiration of the three month period and in particular, that the DTI has a clear record of having received the application within that time (see the discussion in Chapter 10). If the application has been made towards the end of the three month period, then the responsibility is fairly and squarely on the shoulders of the licence-holder to make sure that the application is indeed received by the Department of Trade and Industry in time, even if it may not be acknowledged immediately.

It is the Department's view that the Secretary of State can only grant or refuse the actual necessary wayleave applied for. According to this view, the Secretary of State does not have power to vary the application by, for example, granting a necessary wayleave for one of the electric lines on the owner's land whilst refusing another within the application, on the grounds that it could be better placed elsewhere or underground. It must be at least arguable that this view does not accord with the decision in *British Waterways Board* v *London Power Networks plc*.[6] In that case there was a complaint that the land owner could not influence the route of the wayleave. The Judge explicitly stated that paragraph 6(5) of schedule 4 to the Act gave the Secretary of State "power to control the route" and power to impose conditions "to ameliorate the effect of the wayleave on the landowner or occupier". However until this point is clarified by a further decision or by additional legislation (as to which see Chapter 22), the pragmatic advice is to operate within the ambit of the Department's view. A licence-holder should, therefore, consider whether electric lines operating at different voltages are to be the subject-matter of a necessary wayleave application, and whether it would be preferable to make applications for separate necessary wayleaves for the lines in question.

The view taken by the Secretary of State on the advice of his officials will also be very important in considering to what extent a necessary

---

6   [2003] 1 All ER 187 considered in Chapter 8, section 8.3 on another point.

wayleave application has reasonable prospects of being resisted. This will be discussed in more detail when we consider the scope of the phrase "necessary and expedient" in Chapter 10.

9.13 Stage 4

Assuming that all of the above matters have been satisfactorily managed, that brings Stage 3 to an end. Stage 4 is the DTI's initial response to the application, namely checking that the Secretary of State has jurisdiction to entertain the application. The Department states that it will normally confirm within two months whether or not the Secretary of State has jurisdiction and in the meantime, it may request further information to assist it with that decision. Jurisdiction depends upon two matters. The first is that the application is made in time. The second is that the application is not ruled out by the "dwellings exception". The dwellings exception means (paragraph 6(4) of schedule 4 to the Act) that the Secretary of State cannot grant a necessary wayleave in any case where:

- the land is covered by a dwelling, or will be so covered on the assumption that any planning permission which is in force is acted on and
- the electric line is to be installed on or over that land.

This may therefore require investigation as to whether the inspector is able to consider the application for a necessary wayleave in circumstances where an overhead electric line is to cross land included in these definitions. Clearly a residential property with an immediately attached garden would readily fall within this definition. Significant emphasis was placed on the routing of overhead electric lines over gardens in early 20th century electricity legislation and this has been extended in the Act. Great care must be taken by a licence-holder in proposing an overhead line across land that may be considered to be part of or attached to a dwelling or where a residential planning consent exists for the land. The ability to survey such land, in order to appraise its suitability for an electric line, may also fall within the definition of buildings, considered later in this chapter, where the licence-holder would be unable to rely on their statutory rights of entry.

The DTI have procedural guidance to assist inspectors in carrying out tests to establish whether the land concerned conforms to the definition of dwelling. It should also be noted that the definition of dwelling includes appurtenances for which there is no definition within the Act nor is there a definition of garden. These tests are considered in Chapter 10.

9.14 Lopping and felling of trees

Rights under both statutory and non-statutory wayleaves include rights of access for the maintenance of an electric line. However where the electric line is affected by trees growing in the vicinity of the line, the rights of licence-holders to carry out the felling and lopping of trees in proximity to electric lines is provided for in paragraph 9 of schedule 4 of the Act. This sets out the procedure for serving notice on the owner of the offending trees requiring him to cut them.[7] It should be noted that this is not a notice stipulating that the licence-holder will carry out the work, although in reality it will be the licence-holder or their approved contractors that do carry out the work. It follows as a matter of practical consequence that cutting trees in proximity to live lines must be done under experienced supervision, unless approved contractors are employed. In addition to trees interfering with the operation of an electric line there is also provision for cutting trees where it would constitute an unacceptable source of danger (whether to children or to other persons).[8] The notice from the licence-holder requires, within 21 days from the giving of a notice, that the trees are felled or lopped or the roots cut so as to not obstruct the working of the electric line or constitute an unacceptable source of danger. If within 21 days the owner of the trees serves a counternotice objecting to the requirements of the notice then, in the event of the counternotice not being withdrawn, and no agreement reached between the tree owner and the licence-holder, then it should be referred to the Secretary of State by the licence-holder for the rights to be granted in place of the owners' permission. Accordingly arrangements must be made to convene a hearing at a venue in the locality of the trees concerned. This is similar to a wayleave hearing as the matter still remains effectively private between the tree owner and the licence-holder. Similar formalities are required and the process follows the procedure for

---

7   Paragraph 9 (2) of schedule 9 to the Act.
8   Paragraph 9(1)(b) of schedule 9 to the Act.

necessary wayleaves discussed in Chapter 10. The debate to be had however is more localised and the impact of the line in general is not at issue. If the tree owners' objection is to the line rather than to the impact of the tree cutting required they would be better advised to pursue the wayleave termination process in order that wider issues can be debated on the need and expediency for the line.

Both the owner and the licence-holder are then given an opportunity of being heard by an Inspector appointed by the Secretary of State. Following that the Secretary of State may make an order as he thinks just to empower the licence-holder to cause the trees to be felled or lopped or have the roots cut, having given notice of that to the person serving the counter notice, and to determine any costs to be paid.

Assuming a consent to cut the trees is granted by the Secretary of State the licence-holder is required to carry out the work in accordance with good arboricultural practice and to do as little damage as possible to trees, fences, hedges and growing crops and to follow the directions of the owner in respect of the removal of the felled trees, lopped boughs and root cuttings, in addition to making good any damage to the land. In this context the definition of tree includes any shrub and references to felling or lopping, felled trees or lopped boughs shall be construed accordingly.[9] The requirement for tree cutting mainly applies to overhead electric lines, but can apply to the need to cut the roots of trees in proximity to underground cables.

It is also possible for the tree to be growing on adjacent land to the electric line. The definition of land in paragraph 9 of schedule 4 of the Act,[10] in relation to the serving of a notice on the owner/occupier, relates to the land on which the tree is growing. Therefore an owner/occupier may not be subject to the terms of a wayleave for the electric line if the line is affected by trees growing on adjacent land. In that situation if the tree owner objects to the tree cutting they would have to argue that point alone. If the concerns of the tree owner amounted to the method of the tree being cut they would be entitled to appoint their own contractors to carry out the tree cutting and recover the costs of so doing from the licence-holder. In practice the work would be carried out in conjunction, and by arrangement with, the licence-holder to ensure safe working arrangements were carried out in proximity to the electric line or for taking the line out of commission

---

9   Paragraph 9(8) of schedule 9 to the Act.
10 Paragraph 9(1) of schedule 9 to the Act.

during the tree cutting operations. In addition to recovering these costs the tree owner is also entitled to claim for the loss in value of the trees cut. This would extend beyond the timber value of the trees and include any loss in aesthetic and amenity terms provided by the trees.

At this stage the environmental obligations of the licence-holder under schedule 9 of the Act need to be considered. Under schedule 9[11] the licence-holder is required to take into account the effect on flora and fauna of their proposals and to mitigate the effect of their proposals as follows.

> In formulating any relevant proposals, a licence-holder or a person authorised by exemption to generate or supply electricity —
>
> (a) shall have regard to the desirability of preserving natural beauty, of conserving flora, fauna and geological or physiographical features of special interest and of protecting sites, buildings and objects of architectural, historic or archaeological interest; and
>
> (b) shall do what he reasonably can to mitigate any effect which the proposals would have on the natural beauty of the countryside or on any such flora, fauna, features, sites buildings or objects.

These rights were considered in Chapter 5. Essentially the licence-holder would need to demonstrate to the inspector that the extent of tree cutting complies with these obligations balanced with compliance with the safety requirements of installing or maintaining the electric line or constitute an unacceptable source of danger.

9.15 Entry for survey by licence-holders

Where a licence-holder needs to enter land for the purposes of carrying out a survey on an existing line then their rights are covered by the express or implied provisions of the wayleave for the electric line across that section of land.

Where a survey is required to determine the suitability or practicality of a proposed line any person can carry out a survey on private land where authorised in writing by a licence-holder at any reasonable time provided 14 days' notice or the intended entry has

---

11   Paragraph 1(1) schedule 9 to the Act.

been given to the occupier and that person carries written evidence of his authority.[12] Any person who intentionally obstructs a person acting in the exercise of these powers shall be liable to a fine.[13] Any damage to land or moveables suffered as a result of the survey is to be made good or compensation paid.[14] Any disputes over compensation are to be referred to the Lands Tribunal as discussed in Chapter 16. However just as there is a limitation on the ability to apply for a necessary wayleave where dwellings are concerned similar restrictions apply to the ability to survey land. The powers provided by paragraph 10 of schedule 4 to the Act to carry out a survey do not extend to any land covered by a building, the definition of which including garden, yard, outhouse or appurtenances belonging to or usually enjoyed with a building[15] or on land that has the benefit of planning consent for such purposes.[16] In those cases an agreement would need to be reached with the owner/occupier or the compulsory purchase order proceedings, where notice of entry can be given following the procedures laid down in schedule 3 to the Act for compulsory acquisition for the right, can be utilised. This potential is discussed further in Chapter 11.

9.16 Invasive surveys

Where it is necessary to carry out investigations into the nature of the ground and subsoil to determine their suitability for the proposal, 14 days notice must be given of the proposed works by the licence-holder. There is specific provision that where the survey is to be carried out on land held by statutory undertakers, who would also be served with notice, and they object to the works then the work could only commence with the consent of the Secretary of State.[17] It will be noted that where land is covered by buildings, or will be so covered on the assumption that a planning permission is acted upon; the licence-holder will not be not able to exercise powers to survey or investigate the land under these provisions. That being the case the licence-holder would have to utilise the compulsory acquisition procedures provided

---

12 Paragraph 10 of schedule 4 to the Act.
13 Paragraph 11(1) of schedule 9 to the Act.
14 Paragraph 11(2) of schedule 9 to the Act.
15 Paragraph 10(6) of schedule 9 to the Act.
16 Paragraph 10(3) of schedule 4 to the Act.
17 Paragraph 10(4)(b) of schedule 4 to the Act.

for under schedule 3 to the Act as discussed above. The situation could potentially create difficulties for the licence-holder where insufficient information is available to determine the route or scope of their proposed works for a proposed electric line.

9.17 Survey by inspector

There is one further category of rights to survey and inspect land in addition to the categories identified above. Where an application has been submitted by a licence-holder for a necessary wayleave, in addition to the powers of survey for licence-holders, any land to be inspected by an appointed inspector is empowered by the 1967 Rules to enter the land concerned, without notice, before, during or after the hearing. Clearly this right does not extend to the ability to carry out invasive exploration, such as that required to determine the nature of the subsoil. In entering private land the inspector may find himself in a conflict situation in that good practice would require that notice of a site visit is served on the landowner but any contact with one side and not the other where quasi-judicial proceedings are in progress could amount to procedural unfairness. In those situations it is considered likely that the inspector would make a discreet visit to the property without entering the land and where it was necessary to carry out an inspection of the land concerned then this would be announced during the hearing with the time and date made known to both parties. Given such notice, if one party failed to attend, the inspector would still be entitled to complete his inspection.

Separate provision is made for an appointed inspector, in relation to a necessary wayleave application, both prior to the hearing and following the close of the wayleave hearing, where the inspector can enter the land for the purposes of surveying the area without requiring explicit approval of the landowner/occupier as provided for by the 1967 Rules. This is discussed further in Chapter 10.

Necessary Wayleave Hearings

10.1 Introduction

The factual and statutory context of this chapter is that a licence-holder has applied to the Secretary of State for the grant of a necessary wayleave using the provisions at paragraph 6(3) of schedule 4 to the Act in the circumstances contemplated by paragraph 6(1) and 6(2) of schedule 4. Those were discussed in Chapter 9. Paragraph 6(5) of schedule 4 provides that before granting a necessary wayleave in response to such an application, the Secretary of State "shall afford (a) the occupier of the land; and (b) where the occupier is not also the owner of the land, the owner, an opportunity of being heard by a person appointed by the Secretary of State".

The hearing at which such an opportunity is "afforded" is called the "necessary wayleave hearing".

10.2 Outline of chapter

This chapter explains how in practical terms a date for a necessary wayleave hearing is fixed and secured in the diary. It also explains the rules applicable to the preparation for and the conduct of a wayleave hearing. Further, this chapter considers the fundamental issue which the necessary wayleave hearing has to address. Finally, it explains what happens after a wayleave hearing in terms of notification of the outcome.

10.3 Applicable statutory rules

The present legislation specifically governing necessary wayleave hearings (apart from schedule 4 to the Act itself) are the Electricity (Compulsory Wayleaves) (Hearings Procedure) Rules 1967[1] ("the Rules"). As explained in Chapter 6[2] "compulsory" means "necessary" as defined by the Act. In this chapter, references to a specific Rule will be by use of the following abbreviation: 1967 Rules [Number]. 1967 Rules 1–3 inclusive are the usual formalities to be found in all statutory instruments. We are concerned in this chapter with 1967 Rules 4–11 inclusive.

The rules came into force on 17 April 1967 and obviously in their original form referred to legislation which existed before the Act. The rules were updated by the Electricity Act 1989 (Consequential Modifications of Subordinate Legislation) Order 1990 so that all the statutory cross-references were modernised and harmonised with the Act. In this chapter, citations from the rules are citations from the rules as so modified. Although we have throughout this book generally referred to the grantor of a wayleave as the landowner or occupier, the rules describe the person opposing the grant of a necessary wayleave as the objector and that is the term we shall use in this chapter.

The rules have been extensively supplemented by the DTI guidance to which we have already referred.

10.4 Administration of the application by the Department of Trade and Industry

On receipt by the Secretary of State, an application for his consent under paragraph 6(3) of schedule 4 to the Act to the placing of an electric line across land, a date and time and place for the hearing shall be fixed.[3] This may indicate that the making of an appointment for the necessary wayleave hearing is an automatic next step after completion of Stage 5. This is not the case.

Certainly the DTI recognises that on receipt of a valid necessary wayleave application, the Secretary of State is obliged to offer the owner and/or occupier the opportunity of being heard by an appointed

---

1 SI No 450 of 1967.
2 Chapter 6, section 6.2.
3 1967 Rules 4(1).

person in accordance with the Act before deciding whether or not to grant a necessary wayleave. However, it also recognises that in most cases, it is usual for the licence-holder to carry on negotiations with the owner and/or occupier in order to try and reach an amicable settlement without the need for a hearing.[4]

It is in fact very rare for a hearing to be fixed immediately. The DTI will postpone the appointment for as long as either party indicate that negotiations are continuing. It is important at this stage to note that the rules make no provision whatsoever for the period between the completion of Stage 4 and the actual fixing of a hearing date. There are no longstop dates. Nor is there any requirement that a request for a postponement be justified. Most crucial of all, the DTI does not case-manage the application. Consequently, whereas in an action brought in court or to a lesser extent in an arbitration (when there will be formal rules governing the way the parties communicate with each other and with the court/arbitrator and there will also be regular points at which the parties must report back to the court/arbitrator) there is nothing of the kind governing a necessary wayleave application. The DTI does not issue directions. The DTI prefers but cannot insist upon all communications to it being copied to the other side. As long as all the parties indicate that negotiations are live, the necessary wayleave application will be put into hibernation. That hibernation can be for months and in some cases years.

Why would either party tolerate that? The answer is undoubtedly, first, that there is no provision under schedule 4 to the Act or under the Rules for the reimbursement of the costs to the parties.[5] Naturally, therefore, the parties will only wish to incur the costs of a necessary wayleave hearing if it is indeed (without any sense of irony) really necessary. Second, the inspector cannot award compensation.[6] The DTI makes clear that questions of compensation payments in respect of a necessary wayleave will not be addressed by the inspector of the necessary wayleave hearing (although issues which relate to the impact on the use or enjoyment of the land and may subsequently be subject to a claim for compensation may be raised in evidence at the hearing). The Secretary of State has no power under Schedule 4 to the Act to prescribe financial conditions in any necessary wayleave he may grant or to resolve disputes on the level of compensation. Compensation will

---

4   DTI (4.1).
5   DTI (4.4) and DTI (4.18).
6   DTI (4.17).

fall to be settled by agreement between the parties or, failing agreement, by the Lands Tribunal at the request of either party.[7]

A further regularly encountered reason is that the real issue is often not so much whether a wayleave is necessary for the particular electric line which is the subject matter of the application, but rather whether the proposed electric line can be undergrounded or diverted and if so, at whose expense. Most negotiations focus upon that issue. Whether a necessary wayleave hearing will be of any practical benefit will often depend upon the extent to which that issue will be resolved in consequence of the inspector's report. At this point, it is important to recall that the Secretary of State does not regard himself as permitted to vary the application by making under grounding a condition of the grant of a necessary wayleave. We discuss the significance of this in more detail below.

10.5 Appointment of the inspector

To continue with our explanation of the process, assuming that either party requests the DTI in writing to fix a hearing, then the DTI will act. The Secretary of State will first appoint a person as an inspector. Note that this does not occur until the request in writing is received. This is because there is no provision under schedule 4 to the Act for the DTI to recover the costs of the inspector.[8]

The rules, assuming such an appointment then go on to oblige the Secretary of State

> to give not less than 21 days notice in writing of such date, time and place to every objector and to the licence-holder: provided that:
>
> - With the consent in writing of the objectors and of the licence-holder, the Secretary of State may give such lesser period of notice as may be agreed and in that event he may specify a date for service of the statement referred to in the next following paragraph later than the date prescribed in that paragraph;
> - Where it becomes necessary or desirable to vary the time or place fixed for the hearing, the Secretary of State shall give such notice of the variations which may appear to him to be reasonable in the circumstances"[9]

---

7   Paragraph 7(4) of schedule 4 to the Act.
8   DTI (4.4).
9   1967 Rules 4(1).

10.6 Pre-hearing meeting

It might be thought from that that there will be one single hearing which might be adjourned from time to time. In fact, the DTI interprets this as enabling (but not requiring) both a pre-hearing if the inspector so wishes,[10] as well as requiring the main event itself to take place. The DTI clearly favours a pre-hearing meeting[11] and one is invariably held.

What therefore happens when a party requests the fixing of a necessary wayleave hearing is that a pre-hearing meeting is fixed. Parties will be consulted of possible dates for a pre-hearing meeting.[12] While the DTI will try to meet the wishes of both parties on timing, the Secretary of State's responsibility to act expeditiously on applications may lead the Department to impose a date if agreement cannot be reached. In circumstances where more than one request has been made for a hearing into necessary wayleave applications relating to the same electric line, the Secretary of State will usually consider it appropriate to hold concurrent hearing.[13]

The DTI always asks the licence-holder to make practical arrangements for both the pre-hearing and the main event. This will include arranging a suitable venue for the hearing near to the location of the existing or proposed electric line and to make arrangements for a verbatim transcript of the proceedings to be taken. The licence-holder should contact the DTI if it has any doubts over the type of venue required or how the venue should be set out for the hearing.[14] The requirement that the venue be near to the location of the existing or proposed electric line means that the customary candidates for a venue will be local council offices, church halls and community centres.

Before turning to what happens at a pre-hearing meeting, we ought to explain that if the parties wish to adjourn the pre-hearing meeting because negotiations are proving fruitful, the DTI requires that both parties must write to it asking for an adjournment on at least seven days' notice.[15] This requirement is prompted by the DTI's desire to avoid wasting costs since those of the inspector cannot be recovered from the parties.

---

10 DTI (4.6).
11 DTI (4.6).
12 DTI (4.7).
13 DTI (4.2).
14 DTI (4.3).
15 DTI (4.4).

The DTI makes clear that if the deadline is missed, the pre-hearing meeting will only be cancelled if the necessary wayleave application is formally withdrawn. This will be of no concern to the land owner/occupier for if the application is withdrawn, the electric line must, in principle, either be removed or not installed. Taking account of the time required to reach the pre-hearing meeting stage, it would be highly unlikely for any cancellation to occur within the three month period from the making of an application.

This will be of considerable concern to the licence-holder. However, the licence-holder would have the option to make a compulsory purchaser order.[16] If the application concerned a wayleave to permit a new line then such a state of affairs would have the potential to disrupt severely the project of which the new line formed part. The disruption would be more immediate and self evident if the application concerned an existing line and the licence-holder's obligation to maintain a part of the electricity system in that area. Accordingly, even if negotiations are progressing well, the licence-holder must have a plan B to cover the pre-hearing meeting against the possibility that the negotiations do not reach a mutually acceptable conclusion. By definition, therefore, if the application is formally withdrawn in order to secure the cancellation of a pre-hearing meeting the licence-holder will have lost the right to make fresh application for a necessary wayleave.

The DTI applies the 21 day rule to the pre-hearing meeting just as it does to the necessary wayleave hearing. However, the question is who is actually entitled to receive notice of it? 1967 Rules 3(i) defines "objector" as an owner or occupier of the land across which consent to place an electric line is sought. Note that it does not include pressure groups, village preservation societies or other interested parties. Hence, in the making of the necessary wayleave application, it is essential to identify all relevant land interests correctly. In this connection, licence-holders need to pay particular care to the way in which relevant objectors are identified by reference to a plan. The recommendation of the DTI is that the application should clearly identify the land which is the subject of the necessary wayleave application and the name of the landowner (and where the landowner is not also the occupier) the occupier of the land. Applications should also attach a map clearly showing the boundaries of the owner and/or occupier's land affected and the route of the proposed electric line across the land, including the

---

16   See Chapter 11.

position of any supporting poles or towers. Many land interests can be identified from the Land Registry. However, even today, not all land is enjoyed by way of a registered title. Local enquiries will be essential. Moreover, we must remember the general boundaries rule. Land Registry plans and certainly plans attached to leases or other documents are only accurate to the limitations of their scale. In respect of plans depicting land in rural areas, where a scale of 1/2500 would be utilised, a line drawn thickly on a plan could easily represent 15 m in width. This could therefore make the route of an electric line appear to cross a wider area and, in some cases, give the impression of crossing adjacent land. Licence-holders must be prepared for the fact that some people claim the right to appear both at the pre-hearing meeting and the main necessary wayleave hearing on the basis they own a small strip of land which is nevertheless affected, even though it may not immediately appear from all the usual searches.

10.7 Agenda for pre-hearing

Assuming that the inspector does wish to hold a pre-hearing meeting between the parties (that is to say the licence-holder and those falling within the definition of objector) then it will be held. The main purpose of the pre-hearing meeting is to set out the issues that are likely to be relevant to the Secretary of State's consideration of the necessary wayleave application and agree a provisional timetable for the actual hearing. The principal issue in all such applications is discussed in detail in section 10.15.

10.8 Pre-hearing meeting

A pre-hearing meeting also provides an opportunity for the inspector and the parties to agree a timescale for the exchange of proofs of evidence and any associated documents in order that the hearing may be conducted efficiently and effectively in the interests of all parties.[17] The crucial word in the preceding sentence is agree. Because the rules make no provision specifically for a pre-hearing meeting or indeed stipulate the period between the making of the application and the holding of the actual necessary wayleave hearing itself, there are no

---

17   DTI (4.6).

directions relating to the matters that may be discussed at a pre-hearing meeting. While a timetable may be agreed between the licence-holder and objectors, there is no provision enabling either party to seek a sanction from the inspector if the timetable is not adhered to. In particular, unlike a legal action proceeding in court and/or an arbitration, there is no provision equivalent to the ability to make an application for a claim to be struck out if directions are disobeyed or a claim is inappropriately presented.

It is, of course, in the interest of all parties to make an agreement as to directions, bearing in mind that if a hearing is extended because of inefficient preparation, that impacts upon both the licence-holder and the objectors given that neither can claim costs against the other or from central funds.[18]

Practical matters which will invariably be discussed at a pre-hearing meeting will include some or all of the following:

- agreement of a single plan showing all relevant ownerships and the location of the line including any necessary supports. This ought to be on a large scale so that it can be permanently displayed throughout the necessary wayleave hearing
- the exchange of any expert evidence relating to the need for the line and its impact. This will include not only specifically electrical expert evidence, but also it may include valuation evidence and an environmental impact assessment
- if the title to any of the objectors is challenged, then proof of title ought to be brought into consideration. Indeed, it might well be necessary to invite the inspector to consider such a matter as a preliminary issue so that at the actual necessary wayleave hearing itself, only those genuinely entitled to be there are present
- provision for transcript writers to be present and circulation of transcripts
- the timing of the site inspection.

10.9 Directions imposed by statutory rules

The one specific direction which the Rules do impose is that not later than 14 days before the date of the necessary wayleave hearing (except where the Secretary of State specifies a later date) the licence-holder

---

18   DTI (4.18) and *R v DTI, ex parte Healaugh Farms* The Times 27 December 1995.

shall, unless it has already done so, serve on each objector a written statement of its reasons for the proposed placing of the electric line and shall supply a copy of the statement to the Secretary of State.[19] The rules amplify this by providing[20] that where the licence-holder intends to refer to or put in evidence at the hearing documents (including maps, photographs and plans), the licence-holder's statement shall, unless the licence-holder has already furnished each objector with copies of such documents, be accompanied by a list of such documents, together with a notice stating the times and place to which the documents may be inspected by any objector and the licence-holder shall afford every objector a reasonable opportunity to inspect and where practicable to take copies of the documents.

The DTI amplifies these rules by suggesting that it is helpful for the licence-holder to supply copies of the documents specified, as well as the list to the objector and the inspector at the same time it serves its statement of reasons. Indeed, from the point of view of the efficient conduct of the necessary wayleaves hearing and to the matter of prudent public relations, that is certainly recommended best practice.

On the question of evidence, the Rules make one further point which is not picked up in the DTI guidance. Where a government department has expressed in writing to the licence-holder a view in support of the proposed placing of the electric line and the licence-holder proposes to rely on such expression of view in its submissions at the hearing, the licence-holder shall include that statement of support in the written statement of reasons for the proposed placing of the electric line which has already been mentioned. It follows that the objectors will be able to see what the government department has said in support of the application.[21]

10.10 Submissions by objector

The rules do not in fact place any requirement on the objector to serve upon the licence-holder any evidence or submission which the objector may have been intending to bring to the necessary wayleave hearing. Perhaps when the rules were originally drafted, it was not expected that objectors would be willing to challenge to the extent

---

19 1967 Rules 4(2).
20 1967 Rules 4(4).
21 1967 Rules 4(3).

which is now quite conventional. The DTI guidance[22] suggests that if any objector intends to present a lengthy submission, either as a statement of reason or in evidence, it would be helpful if it were presented to the other parties and the inspector in advance of the beginning of hearing.

10.11 Notice of main hearing

As previously mentioned, the DTI will give the parties at least 21 days' notice of the date, time and place of the necessary wayleave hearing unless they agree a lesser period of notice in writing. As previously indicated, the Secretary of State has the discretion to vary the date, time and place of the hearing by giving prior written notice.[23] The Department will therefore normally agree to postpone a hearing already arranged if parties consider that a settlement is likely. In such cases, the written agreement of both the electricity company and the objector is required.[24] Although not specifically expressed, it seems clear that the Department would likewise expect a minimum of seven days' notice.

10.12 Representation at main hearing

The licence-holder may appear at the hearing by any of its officers appointed by it for the purpose or by counsel or by solicitor, and an objector may appear on his own behalf or be represented by counsel, solicitor or any other person.[25]

Where there are two or more objectors having a similar interest in the matter under inquiry, the inspector may allow one or more persons to appear for the benefit of some or all of the objectors so interested.

Where a government department has expressed in writing to the licence-holder a view in support of the proposed placing of the electric line and the licence-holder has set out such a view in the statement referred to, any objector may, not later than seven days before the date of the hearing, apply in writing to the Secretary of State for a

---

22 DTI (4.9)
23 1967 Rules 4(1).
24 DTI (4.10).
25 1967 Rules 5(1).

representative of the government department concerned to be made available at the hearing.[26]

The Secretary of State shall transmit any application made to him under such a request to the government department concerned who shall make a representative of the department available to attend the hearing.[27]

Such a representative shall at the hearing state the reasons for the view expressed by his department and shall give evidence and be subject to cross-examination of the statement of other witnesses, so, however, that the appointed person shall disallow any questions which in his opinion are directed to the merits of government policy.

10.13 Conduct of the hearing

The inspector has complete control over the procedure to be adopted at the hearing. Everything is within his discretion.[28] As the DTI points out[29] it should be noted that a necessary wayleave hearing is not a public inquiry where members of the public may also contribute. This means that only the licence-holder and those properly qualified as objectors (owners and occupiers or their representatives) are entitled to appear and make their case. They may of course call expert or relevant witnesses to give evidence on their behalf. However, it should also be understood that the parties may ask for the hearing to be conducted in private and if so, it will be held in private.[30]

Although generally the inspector does have complete discretion about the running of the necessary wayleave hearing, a number of points are mandatory. Unless in any particular case the inspector with the consent of the licence-holder otherwise determines, the licence-holder shall begin and have the right of final reply and the objectors shall be heard in such order as the inspector may determine. The licence-holder and the objector shall be entitled to call evidence and cross-examine persons giving evidence. The appointed person shall not require or permit the giving or production of any evidence whether written or oral which would be contrary to the public interest, but

---

26   1967 Rules 6(1).
27   1967 Rules 6(2).
28   1967 Rules 7(1).
29   DTI (4.12).
30   1967 Rules 7(2).

apart from that exception and without any way undermining the ban on questioning of government policy at such a hearing, any evidence may be admitted at the discretion of the appointed person who may direct that the documents attended in evidence may be inspected by any person entitled to appear at the hearing and that facilities be afforded to that person to take or obtain copies of those documents.[31]

It is to be noted that the DTI guidance[32] merely says "the necessary wayleave hearing proceeds by way of presentation of evidence, cross-examination and re-examination of those giving evidence". What the guidance does not go on to say is that under the rules, the inspector may allow the licence-holder both to alter or add to the reasons contained in the written statement served 14 days beforehand, or to alter or add to any list of the documents which accompany it, so far as may be necessary for the purposes of determining the questions being debated between the parties. However, although the licence-holder does have the freedom to expand and alter the case in this way and produce additional documentation, it cannot surprise any of the objectors. The rules goes on to provide that if the inspector does permit the licence-holder to take advantage of this privilege, the inspector shall (if necessary by adjourning the hearing) give every objector an adequate opportunity of considering any such alterations or additions.[33]

There is a general power to adjourn the hearing from time to time and if the date, time and place of the adjourned hearing are announced before the adjournment, no further notice shall be required.[34]

If any objector does not appear at the hearing, the inspector may at his discretion proceed with the hearing and if the inspector does so, the inspector shall (subject to making disclosure of those facts at the hearing) take into account any previous written representations of the objector insofar as the same appear to him to be proper and relevant to the matters and issues. Note that an objector who fails to appear at the necessary wayleave hearing either in person or by some representative cannot, on the basis of his absence, ask for the hearing to be re-listed. While the absence of any objector might be regarded as an advantage to the licence-holder, this is not necessarily the case. Bearing in mind that any statement can be taken into account, the absence of the objector would mean that the statement is not contested by cross-examination. If

---

31   1967 Rules 7(3) to 7(5) inclusive.
32   DTI 4.13.
33   1967 Rules 7(6).
34   1967 Rules 7(8).

the statement contains controversial statements as to the facts, then this actually could be a disadvantage to the licence-holder. This underlines the importance of the licence-holder when making the application in the very first place, to have correctly identified all the relevant land interests.

10.14 Site inspection

If he has not already done so prior to or during the hearing, then following the closure of the evidence, the inspector will generally inspect land affected by the electric line accompanied by representatives of the licence-holder and the owner and/or occupier. The formal inspection, which is designed to give parties the opportunity to point out features addressed at the hearing, is not, however, automatic. The rules merely give the inspector the discretion to make an unaccompanied inspection of the land before, during or after the hearing without giving notice of his intention to any person entitled to appear at the hearing.[35]

A party who takes the view that an inspection is required ought to make a formal request to that effect to the inspector either before or during the hearing. The rules do state that the inspector shall, if so requested by the licence-holder or any objectors before or during the hearing, inspect the land after the close of the hearing and shall, when such a request is made, announce during the hearing the date and time which it proposes to make such an inspection.[36] Although the licence-holder and the objectors (or as supplemented by the DTI guidance, their representatives) shall be entitled to accompany the inspector on any inspection held as a result of a request made, the inspector shall not be bound to defer his inspection if any person entitled to accompany him is not present at the time appointed. There seems to be little advantage in such a deferment in any event since if the inspection is occurring after the hearing, it is important that it occurs while the evidence is still fresh in the inspector's mind.[37]

The DTI goes on to state that no further evidence may be given or points raised during any formal site inspection (the inspector cannot refer to any such evidence or points in his report or take them into account in his considerations).[38] However, it is difficult to see why the

---

35 1967 Rules 8(1).
36 1967 Rules 8(2).
37 1967 Rules 8(3).
38 DTI 4.13.

DTI makes this point. There appears to be nothing specific in the rules requiring the inspector to restrict the scope of the inspection in this way. Indeed, if all parties are represented either in person or by their professional representatives, then nobody suffers any prejudice if one point is made as long as the opportunity to respond is also given. Indeed, even pointing out the correlation between the physical features on the land and evidence that may have been given in the hearing would seem to be the giving of evidence to some degree and so inevitably there will have to be some latitude in that respect, bearing in mind that otherwise the inspection has to occur in complete silence.

At this point, the rules go on to discuss the procedure after the hearing. We pause in consideration of the rules because it is now appropriate to consider the main question the inspector must address in his recommendations: namely whether the wayleave is necessary and expedient for the purposes of the proposed electric line.

10.15 Necessary and expedient

The Oxford English Dictionary defines these two words as follows:

- Necessary — 1. Indispensable, requisite, requiring to, that must, be done. 2. determined by predestination or natural laws, not by free will; happening or existing by necessity (necessary evil); (of concept or mental process) inevitably resulting from nature of things or the mind, inevitably produced by previous state of things; having no independent volition. 3. Thing without which life cannot be maintained (the necessaries of life) or is unduly harsh; money or action needed for a purpose.

- "Expedient — Advantageous, suitable, (do whatever is expedient); politic rather than just. Resource, means of attaining one's end.

The DTI guidance[39] states that the purpose of a necessary wayleave hearing is to hear evidence as to:

- why it is necessary or expedient for the electric line to cross the particular land in question and
- what the effects are of the electric line on the use and enjoyment of the land in question.

---

39 DTI (4.14).

The effects will be a matter of fact combined with an expert opinion. More difficult is the exact scope of the phrase "necessary or expedient". There is a limit, for instance, to the extent to which questions of a more general nature such as the overall case for a new electric line or the retention of an existing one can be dealt with at the hearing. Such matters are more properly considered if the licence-holder has applied for a consent under section 37 of the Act. However, while it will not always be necessary for the electric line to go over a particular set of landholdings because electricity can be supplied in a different way, it may equally not be expedient for the line to follow a particular direction. What the inspector has to focus on is establishing the effect on the private land interests of those whose land lies under the existing line or would be under the new line if installed. The evidence that will be relevant at a necessary wayleave hearing has to be site specific, for example the effect of the actual line in question on a farm (crops and animals) on the use of machinery, or flora and fauna and in the case of an overhead line, on the outlook from buildings situated on the land in question. Other relevant evidence is likely to be the cost of any suggestions for local diversions of the application route (typically up to a maximum of 500 m either side of the existing/proposed route) and in the case of an overhead line, the location and supports on the land in question.[40]

The crucial question is the extent to which the inspector can deem an electric line, either existing or new, neither necessary nor expedient on the basis that an undergrounding option is available. At this point we need to bear in mind the DTI's guidance which says that the Secretary of State cannot approve an application conditionally by requiring part of the overhead line to be installed underground. Accordingly, if the licence-holder has not applied for a necessary wayleave for an underground cable, then the objectors have to be confident that they can beat the application entirely. They cannot as it were negotiate a revised form of application through the medium of the necessary wayleave hearing.

It is for the inspector to provide confirmation that both tests are satisfied. However the balance between the two will need to be considered to show that it is both necessary and expedient for the line in question to be placed or retained on the subject land. Necessity is usually the easier of the two to prove as if an existing line were taken

---

40   DTI (14.15) and (14.16).

away the licence-holder should be able to demonstrate that supplies would be lost or that the licence-holder would be unable to comply with its licence obligations in complying with laid down standards of security of supply. In terms of a line crossing the particular land of an objector the expediency for that route would need to be established taking account of, local conditions and the consideration of practical alternatives, including their cost. The presence of buildings, existing or proposed, ground conditions, topography, location of watercourses, presence of trees, flora and fauna would all need to be considered. The impact on the use and enjoyment of the land will also need to be considered, including financial loss, but not the award of compensation as this is outside the remit of the inspector.

10.16 Post hearing procedure

After the hearing the inspector makes a report of the hearing with his recommendations and conclusions. The DTI normally allows the inspector three months in which to write it. The report has to include the inspector's findings of fact and his recommendations, if any, or his reasons for not making any recommendations.[41] We stress again that the inspector can only recommend the grant or refusal of the actual wayleave for which application has been made. He cannot impose any conditions or variations with a view to giving weight to the arguments made by an objector.

The Secretary of State then considers the inspector's report and notifies the parties of his decision (normally within two months of receiving the report). Where the Secretary of State differs from the inspector on a finding of fact or after the close of hearing receives any new evidence (including expert opinion on the matter of fact) or takes into consideration any new issue of fact (not being a matter of government policy) which was not raised at the hearing, and by reason of those matters is disposed to disagree with the recommendation made by the inspector, the Secretary of State will not come to the decision which is at variance with any such recommendation without first notifying the licence-holder and any objector who appeared at the hearing of his disagreement and the reasons for it and giving them an opportunity of making representations in writing within 21 days or (if the Secretary of State has received new evidence or taken into

---

41 1967 Rules 9(1).

consideration any new issue of fact not being a matter of government policy) of asking within 21 days of the re-opening of the hearing.[42]

The Secretary of State may in any case if he thinks fit cause the hearing to be re-opened and shall cause it to be re-opened if asked to in accordance with the previous provision. If the hearing is re-opened, the same procedure shall apply as if we were fixing the original hearing.[43]

The Secretary of State notifies his decision and his reasons for the decision in writing to the licence-holder and to the objectors and where a copy of the inspector's report is not sent with the notification of the decision, the notification shall be accompanied by a summary of the inspector's conclusions and recommendations. The DTI guidance in fact states that a copy of the inspector's report will be included and if a necessary wayleave is being granted, the wayleave document itself will be sent.[44]

If after considering the inspector's report and recommendations, the Secretary of State decides to grant the necessary wayleave, it will usually be granted subject to a condition that it may only be terminated after 15 years. Remember, at this point, that necessary wayleaves granted by the Secretary of State are binding on successive owners and/or occupiers.[45] It is the DTI's considered view that a 15 year term represents an equitable period which provides a balance between offering the licence-holder a degree of certainty for the installation of the electric line while still affording the landowner the opportunity of having the position reviewed in the light of subsequent changes and circumstances on the local environment. Accordingly, parties who consider that granting a necessary wayleave over a shorter or a longer term would be more appropriate because of the particular circumstances, should give their reasons as part of their evidence.[46]

The necessary wayleave will have to define the land over which it is to run and the physical route of the electric line. In consequence, the necessary wayleave document will invariably have annexed to it an appropriate plan. The DTI's policy therefore makes it unnecessary for any objector in reality to resort to 1967 Rules 10(2) and 10(3) which provide the ability for an objector to obtain a copy of the report on application to the Secretary of State and obtain inspection of documents, maps, photographs and plans referred to.

---

42 1967 Rules 9(2).
43 1967 Rules 9(3).
44 1967 Rules 10(1) and DTI (5.1).
45 Paragraph 6(6)(b) of schedule 4 to the Act.
46 DTI (6.10).

Because this is an executive decision of an arm of government and not the decision of a court or of an arbitrator, there is no right of appeal.

10.17 Documenting a necessary wayleave

If the Secretary of State decides to grant a necessary wayleave then a letter is sent by the Director of Electricity Consents to the parties enclosing a document formally granting the consent by reference to an attached plan. The minimum term and the requisite notice period for subsequent termination will be stated. No other terms or conditions are usually spelled out. The *Craven*[47] decision fills in the obvious blanks. No periodical wayleave payment is mentioned because the landowner is entitled to compensation, either by agreement with the licence-holder or, failing agreement, upon application to the Lands Tribunal.

10.18 Award of costs

It has been argued that where a landowner or occupier appears at a wayleave hearing, they should be awarded costs in the event that they are "successful" in not having a necessary wayleave awarded in favour of the licence-holder or where the inspector considers that the licence-holder has acted unreasonably. In a reversal of the latter circumstances the licence-holder may also wish to make a claim where they consider a land owner or occupier increases their costs associated with a hearing. Equally the DTI is unable to recover its costs from either party in any circumstances. The 1967 Rules do not provide for the award of such costs. In compulsory acquisition matters the award of costs is covered under section 250 of the Local Government Act 1972 on the basis that most compulsory acquisitions are promoted by local authorities. However, section 62 of the Act specifically applied sections 250(2) to 250(5) of the 1972 Act to statutory inquiries and not hearings. This was confirmed in the case of R v *Department of Trade and Industry, ex parte Healaugh Farms*.[48] Under an application under Schedule 8 and section 37 of the Act there would be the ability to seek an award for such costs. The position in respect of necessary wayleave hearings is clearly anomalous.

---

47 Discussed in Chapter 8, section 8.4.
48 QBD reported in The Times 27 December 1995.

Compulsory Purchase Powers

11.1 Outline of chapter

This chapter concentrates on the powers available to licence-holders to acquire land for operational purposes and to acquire rights over land for electric lines whether overhead or underground. Reference is also made to the compensation provisions of relevant Acts and to situations where these compulsory powers would be utilised by licence-holders.

11.2 Legislative background

The principal source of statutory provision for the compulsory acquisition of land and property rights by licence-holders is Schedule 3 to the Act. Schedule 3 is given effect by section 10 of the Act which states[1] that subject to one matter[2] Schedule 3 shall have effect in relation to the holder of a transmission licence and to the extent that his licence so provides in relation to an electricity distributor or any other licence-holder and references in Schedule 3 to licence-holder shall be construed accordingly.

The one matter is that where any provision of Schedule 3 is applied to a licence-holder by his licence, it shall have effect subject to such

---

1 Section 10(1) of the Act.
2 Section 10(2) of the Act.

restrictions, exceptions and conditions as may be included in the licence for the purpose of qualifying that provision as so applied or any power or right conferred by or under it.

Part 1 of Schedule 3 (paragraphs 1–4 inclusive) set out the powers of compulsory acquisition and Part II of Schedule 3 (paragraphs 5–14 inclusive) deals with issues of procedure and compensation.

Paragraph 1(1) of Schedule 3 provides that the Secretary of State may authorise a licence-holder to purchase compulsorily any land required for any purpose connected with carrying on of the activities which it is authorised by its licence to carry on. Land is defined by paragraph 1(2) as including any right over land and the power of the Secretary of State includes power to authorise the acquisition of rights over land by creating new rights as well as acquiring existing ones. Given the nature of the activities of a licence-holder, there are occasions where the total acquisition of the land would be unnecessary and therefore rights, for example, to oversail land with conductors would be sufficient. Those would be situations in which paragraph 1(2) of Schedule 3 would be relevant.

There are special provisions for the exercise of powers as between licence-holders. These can only be exercised with the consent of the authority which in turn is subject to various conditions.[3] There are also provisions relating to the acquisition of open space and allotments.[4] Paragraph 1(4) of Schedule 3 states that where a licence-holder has acquired land using Schedule 3 powers the licence-holder must neither dispose of that land nor grant any interest in or right over it except with the consent of the Authority. One licence-holder may compulsorily acquire the land of another with the consent of the Authority.

11.3 Compulsory acquisition between licence-holders

For the purposes of paragraph 2 of Schedule 3 land has a distinct meaning when applied to compulsory acquisition between licence-holders. First the land must not be "operational".[5] Secondly the land

---

3   Paragraph 1(2) Schedule 3 to the Act.
4   Paragraph 1(3) Schedule 3 to the Act.
5   Paragraph 2 (2)(a) of Schedule 3 to the Act.

must not be ripe for future development for operational purposes.[6] If either apply, the Authority must not grant consent.

11.4 Procedure

Apart from the distinction noted in section 11.3 the procedures follow those applicable to compulsory acquisition of land by acquiring authorities under the Acquisition of Land Act 1981. However the Act provides two specific adaptations of the Compulsory Purchase Act 1965. These adaptations provide for the effect on retained land to be taken into account and to protect the landowner against the severance of a house or garden occasioned by the granting of rights over such areas to a licence-holder. Provision is also made for determination by the Lands Tribunal of the requisite compensation.

The adaptations within the Act[7] of the Compulsory Purchase Act 1965 also make provision for situations where the taking of an interest over part of the land would be prejudicial to certain property interests and therefore the acquiring authority, for electric lines the licence-holder, can be obliged to take the whole property. In the case of the acquisition of land by licence-holders the following variation is provided to section 8 of the 1965 Act:

> No person should be required to grant any right over part only of any house, building or manufactory or of a park or garden belonging to a house, if he is willing to sell the whole of the house, building, manufactory, park or garden unless the Lands Tribunal determines that:

> in the case of a house, building or manufactory, the part over which the right is proposed to be acquired can be made subject to that right without material detriment to the house, building or manufactory: or

> in the case of a park or garden, the part over which the right is proposed to be acquired can be made subject to that right without seriously affecting the amenity or convenience of the house;

> and if the Lands Tribunal so determines, the Tribunal shall award compensation in respect of any loss due to the acquisition of the right, in

---

6   Paragraph 2(2)(b) of Schedule 3 to the Act (subject to extension under paragraph 2(3) of Schedule 3 to the Act).
7   Part II of Schedule 3 to the Act.

addition to its value; and thereupon the party interested shall be required to grant to the acquiring authority that right over the part of the house, building, manufactory, park or garden.

In considering the extent of any material detriment to a house, building or manufactory, or any extent to which the amenity or convenience of a house is affected, the Lands Tribunal shall have regard not only to the right which is to be acquired over the land, but also to any adjoining or adjacent land belonging to the same owner and subject to compulsory purchase.

Therefore if the licence-holder considers that the rights can be taken without material detriment to the house, building or garden then they can apply to the Lands Tribunal for a determination. The Lands Tribunal has jurisdiction to determine if the rights can be acquired without material detriment to the house, building, manufactory, park or garden.

As discussed in Chapters 5 and 9, in addition to situations where it is not possible to identify the ownership of land, there will also be occasions where an application for a necessary wayleave could not be entertained by the Secretary of State. It would therefore be open to the licence-holder to make a compulsory purchase order in those situations to overcome an existing electric line being on the land in a trespass situation. An example would be where an application had not been made for a necessary wayleave within three months of a notice to remove the electric lines being served. In that situation the licence-holder would be open to an action for trespass and could either remove the offending line or restrict their vulnerability to an action by making the application for a compulsory purchase order and, when authorised, the apparatus could then remain.

The procedures on powers of the licence-holder are covered in the Acquisition of Land Act 1981 which also require them to apply the compensation code[8] as discussed in Chapter 15. The 1981 Act provides procedures common to all acquiring authorities. In the case of electric lines the acquiring authority is the licence-holder. By way of contrasting example, for motorway schemes the acquiring authority is the Highways Agency. The use of the 1981 Act avoids the need to repeat the procedure in each enabling legislation for organisations that have recourse to powers of compulsory acquisition. There are also some limited provisions from the Compulsory Act 1965 and the Land

---

8   Section 5 of the Land Compensation Act 1961.

Compensation Act 1973 to take account of but a complete explanation of these is beyond the scope of this book and the purpose here is to note that these powers are available to licence-holders.

11.5 Compensation

In terms of an electric line, acquiring a right over the land would provide for general rights where the works were placed but there would be the effect on other land of the original land owner to be taken into account. Therefore in order to aid the assessment of compensation, section 7 of the Compulsory Purchase Act 1965 is substituted by virtue of the Act[9] with the following:

> In assessing the compensation to be paid by the acquiring authority under this Act regard shall be had not only to the extent (if any) to which the value of the land over which the right is to be acquired is depreciated by the acquisition of the right but also to the damage (if any) to be sustained by the owner of the land by reason of its severance from other land of his, or injuriously affecting that other land by the exercise of the powers conferred by this or the special Act.

11.6 When a licence-holder would use compulsory powers

It is rare for licence-holders to utilise these compulsory purchase powers but there are occasions where it is essential. Certainly in cases where it has not been possible to determine ownership or occupation of land for the purposes of seeking a necessary wayleave then the compulsory purchase powers would be needed in order to fill the gap. In other situations, the exercise of the powers would not be essential but the licence-holder may nevertheless have to resort to them. An example would be where a wayleave is considered to be insufficient and the landowner is unwilling to negotiate an easement or to negotiate the sale of land or an interest in the land on reasonable terms.

Obtaining permanent rights through a compulsory purchase order requires a more complicated procedure than the one appropriate to necessary wayleaves and is therefore only applied in situations where

---

9   Paragraph 8 of Schedule 3 to the Act.

the necessary wayleave procedure cannot be employed. The following are examples of situations where this could apply:

- absent or untraced owners
- rights under schedule 4 to the Act have expired
- beyond the scope of schedule 4 to the Act.

Taking each of these in turn: the obligation of a licence-holder to obtain permission from all landowners and occupiers to install an electric line is a requirement of the Act. If however it is not possible to determine the ownership of land then the compulsory purchase procedure can and has been utilised to overcome this obstacle. Schedule 3 to the Act[10] incorporates the provisions of the Compulsory Purchase Act 1965[11] to enable a licence-holder to have a sufficient interest in the land so as to install an electric line or acquire land for their statutory purposes.

The requirements set out in the Acquisition of Land Act 1981 are followed and require, in the absence of being able to serve notice on the landowner, notices to be clearly displayed on the land in question. The detailed requirements under the compulsory purchase procedure are beyond the scope of this book and will not be considered in detail.

In cases where a notice to remove an electric line has been served and the licence-holder has neither removed the line nor applied for a necessary wayleave within three months, but the line remains across private land, then one alternative would be to apply for a compulsory purchase order for the rights to retain the line across the land. If a compulsory purchase order is confirmed then at that stage the licence-holder would no longer be in trespass in retaining the electric line on the land.

Schedule 4 to the Act precludes the Secretary of State from considering an application where the land to be crossed by an electric line is covered by a dwelling, as defined, as set out at paragraph 6(4) of the Act as follows:

> "The Secretary of State shall not entertain an application under sub paragraph (3) above in any case where:
> (a) the land is covered by a dwelling or will be so covered on the assumption that any planning permission which is in force is acted on; and

---

10　Paragraph 10 of Schedule 3 to the Act.
11　Paragraph 2(3) of Schedule 2 to the Compulsory Purchase Act 1965.

(b) the line is to be installed on or over the land.

It follows therefore that if an overhead line is to be installed across a dwelling, or the garden to a dwelling, as defined, then it will not be possible for the licence-holder to apply to the Secretary of State for a necessary wayleave in those circumstances. Equally where planning permission for residential development has been granted for the land in question that also precludes an application from being made. Apart from excluding the proposal to install the electric line across that land the only options available to the licence-holder are to acquire rights by agreement or to promote a compulsory purchase order.

Where easements are obtained by agreement by the licence-holder from a landowner then the provision of compulsory purchase rights under schedule 3 needs to be borne in mind as this sets out the background to the assessment of compensation which, together with the terms of an easement, will determine the amount of compensation to be paid for the land taken, injurious affection and disturbance. This is discussed further in Chapter 15. Schedule 3 to the Act is also relevant where agreement has been reached by the licence-holder with the landowner on the terms for an easement but the consideration remains to be agreed. In any reference to the Lands Tribunal it would be useful to refer to the terms of Schedule 3 in assessing the quantum of compensation to be paid as this forms the background to the negotiations.

Planning Permission and Ministerial Consent

12.1 Outline of chapter

This chapter covers the practice and procedures relevant to planning permissions and ministerial consents for electric lines. There is an overlap with the procedures applicable to an application for a necessary wayleave, particularly in the case of a new overhead line. It is important to understand how these planning permissions and ministerial consents complement necessary wayleaves granted on application to the Secretary of State.

12.2 Introduction

Under section 37 of the Act consent is required for all new overhead lines, with limited exceptions, such consent to be granted by the relevant minister before they can be constructed. This legislation follows the previously accepted practice that where work of public utilities, under a government department, requires a consent then the local planning authority is effectively consulted with the final decision resting, not with the local planning authority, but with the relevant Secretary of State. With electric lines these principles apply apart from the limited exceptions provided for in the Overhead Line (Exemption) Regulations 1990[1] and for service lines as defined under section 37 of the Act.

---

1   SI 1990 No 2035.

Under section 55 of the Town and Country Planning Act 1990 any development of land requires planning permission. Development means:

> the carrying out of building, engineering, mining or other operations in, on, over or under land, or the making of any material change in the use of any buildings or other land.

Express planning consent is not required for certain activities particularly those within the operational land of licence-holders. The erection of overhead lines is generally only carried out by electricity and telecommunication companies and is therefore a discrete type of development. Electricity development is distinct from telecommunications-related development. With electric lines there is also the factor of high voltage and the consequential potential danger that require further consideration.

Exceptions to the need for express planning consent are provided in the Town and Country Planning (General Permitted Development) Order 1995[2] where provision is made for permitted development of statutory undertakings within Schedule 2, Part 17, Class 4. This is reproduced at Appendix 17.

12.3 Applications for consent for overhead lines

The application process requires the licence-holder to consult the local planning authority. This is carried out under the Department of the Environment Circular 14/90 which sets out the details. This requires an application form to be submitted to the appropriate planning authority. The application is in a standard format and the forms are set out at Appendix 12 and 13 respectively. The distinction between the forms is that for high voltage lines, in excess of 132,000 volts, in addition to reporting on consultation with the district council for the area in which the line is to be installed, provision has to be made for reporting on consultation with the county planning authority (if any). Moreover, as well as providing details of the proposed line, the application requires details of any representations or objections made to the licence-holder

---

2   SI 1995 No 0418.

Planning Permission and Ministerial Consent

and a statement of the licence-holder's compliance with Schedule 9 to the Act in respect of environmental obligations. This requires the licence-holder to demonstrate how they have had regard to the preservation of natural beauty, land and buildings of a special or historic interest as discussed in Chapter 8.

For overhead lines operating at less than a voltage of 132,000 volts, an application only is sufficient, but for lines of 132,000 volts and greater the application needs to be advertised in accordance with the Electricity (Application for Consent) Regulations 1990[3] which require advertisement in at least one newspaper published in the locality of the electric line. Within the Form B procedure is a requirement for the licence-holder to give details of the line together with details of objections lodged with them and a statement on the environmental effect of the proposal. Under the Town and Country Planning (Assessment of Environmental Effects) Regulations 1988,[4] as amended in 1994,[5] in respect of lines of 132,000 volts, where for a length greater than 1 km, it is necessary to have an environmental assessment prepared. This is because it is classed as a Schedule 2 project under the European Council Directive 85/337/EEC as implemented by the Electricity Works (Environmental Assessment) (England and Wales) Regulations 2000.[6]

In making the application to the Secretary of State a formal request has to be made for a direction that planning permission is deemed to be granted. This is by virtue of section 90(2) of The Town and Country Planning Act 1990 as below:

> On granting a consent under section 36 or 37 of the Act in respect of any operation or change of use that constitutes development, the Secretary of State may direct that planning permission for that development and any ancillary development shall be deemed to be granted, subject to such conditions (if any) as may be specified in the direction.

There are also situations where a licence-holder would need to carry out modifications to an electric line which would be at variance with the consent granted for the line. This may include, for example,

---

3   SI 1990 No 455.
4   SI 1988 No 1199.
5   The Town and Country Planning (Assessment of Environmental Effects) (Amendment) Regulations 1994 SI 1994 No 677.
6   SI 2000 No 1927.

increasing the number of conductors of the line or increasing the permitted voltage. The licence-holder would need to follow the same procedures as provided for in Department of Environment Circular 14/90 in submitting a consultation to the planning authority as well as obtaining all wayleaves and/or easements for the new line.

12.4 Objections by planning authority or other persons

Where the relevant local planning authority has objected to the application for consent to build an overhead line, and their objection is not withdrawn, the Secretary of State is required to hold a public inquiry. The procedures for this are discussed in Chapter 13. Where objections have been made by other persons, normally following publication of the application in local newspapers, the Secretary of State "shall consider those objections, together with all other material considerations" as to whether a public inquiry should be held. If the Secretary of State is minded to consider other objections, where none have been made by the relevant planning authority, then there is provision for an informal hearing to be held. Despite there being no statutory instrument defining the rules for such a hearing, provision has been made by the DTI for a hearing to follow similar proceedings as for an inquiry under section 37 of the Act. This includes requiring a statement of case by the licence-holder and by qualifying objectors. In these circumstances it would be for the appointed inspector to determine whether the objectors be confined to their written representations, or whether a hearing is held to ensure that the planning application has been properly and fairly conducted. If there is a hearing the appointed inspector may also determine that an additional inspector sit with him if the issues to be resolved need input from another government department.

12.5 Situations where consent is not required

In order to assist the licence-holder to deal with any unforeseen circumstances when constructing a new overhead line, a tolerance is provided with the Secretary of State's consent on the route of the overhead line. This provides a lateral tolerance which varies with the voltage of the line and is usually requested by the licence-holder. A 200 m tolerance each side of the route of the line would be considered appropriate for high voltage lines. However, once the line has been erected and commissioned then that tolerance is spent and any further work required would relate to the actual location of the line. While consent is granted for the route of the line, and tower and pole positions may be indicated, the consent does not specify the number of supports for which consent is granted. As consent is granted for an electric line all supports are included under that definition. However, once the line is operational then that line becomes the completed development and any further modifications, including additional towers and/or poles would require a further consent and again the details of the exemption regulations would need to be checked. Additionally, there is provision for lines to be diverted on a temporary basis. Limited exemptions are provided where further consent is not required with parameters set on the extent of deviation and the time period for retaining the temporary line. Flowcharts are set out at Appendices 14, 15 and 16 dealing with situations where modifications are required to the route of existing overhead lines and the diversion of permanent and temporary overhead lines. Where major lines are to be modified then different criteria apply depending upon the extent of the line diversion and the location of the line.

There are certain types of development that do not require any consent under section 37. These are covered under the terms of section 37 and the Overhead Line (Exemption) Regulations 1990.[7] The provisions of the exemption regulations can be identified from the flowcharts. The exemptions from a consent under section 37 of the Act are set out below:

(1) Subject to subsection (2) below, an electric line shall not be installed or kept installed above ground except in accordance with a consent granted by the Secretary of State.

(2) Subsection (1) above shall not apply:–

in relation to an electric line which has a nominal voltage not exceeding 20 kilovolts and is used or intended to be used for supplying a single consumer;

---

[7] SI 2035 No 1990.

in relation to so much of an electric line as is or will be within premises in occupation or control of the person responsible for its installation; or
in such other cases as may be prescribed.

12.6 Review of minister's consent

The significant difference with this type of consent is that it is capable of being reviewed, as are all consents whether under section 37 of the Act or under preceding legislation, normally after five years of being granted. The decision on whether to review a consent rests with the Secretary of State although very few consent reviews have been carried out. Of those that have, the majority have resulted in the overhead line remaining despite opposition from, usually, local pressure groups. There have been only a handful of consent reviews undertaken by the Secretary of State and even less publicity attached to them. For reasons not known a consent review was undertaken in 1987, under section 10b to the schedule to the Electric Lighting (Clauses) Act 1899, of a wood pole line at Pagham in West Sussex. Consent had only been granted in 1972 and therefore it is not purely tower lines that are subject to a review although the vast majority invariably will be. In 1993 a review of a consent in respect of approximately 2 km of an overhead steel tower line at Krumlin, Halifax took place. The consent for the line had been granted in 1955 with a new consent granted in 1966 to enable the line to operate under greater capacity. This review followed representations from two landowners affected by the line who were unable to utilise the wayleave termination process as easements had been granted in respect of the line. No doubt due to the exposed nature of the land where the line was situated, at 300 m above sea level, National Grid had previously considered a new section of replacement overhead line to replace the number of towers affected by severe weather conditions. The objections to the line were on health grounds. Being in proximity to dwellings raised concerns about electric and magnetic fields from the line. Part of the reason for not building the replacement overhead line was the objection of the local planning authority to these proposals. The conclusion reached was that despite the concern of hazards to health from the line it was recommended that the consent for the line should not be terminated. In Abergavenny, South Wales a consent granted in 1964 was reviewed in 1997. Prior to that date the line had completed residential development beneath it although only three houses were directly affected by the overhead line. The review was undertaken of approximately 1200 m of overhead line.

Where a landowner has terminated a wayleave and/or given notice to remove an electric line it must also be considered whether a review of the consent under section 37 should be requested. It may be that uses in the general area have significantly changed but the extent of land covered by a terminated wayleave may be proportionately small in relation to the entire overhead line. In those situations it is possible for a landowner to ask for the consent under section 37 to be reviewed. The landowner may apply either directly to the Secretary of State as an individual acting alone or apply with the support of the local planning authority or other interested parties. There are no laid down procedures for determining the circumstances in which this would be justified or considered appropriate. Given the examples outlined, each case would need to be considered on its own merits. One would certainly take into account the change in circumstances since the consent was originally granted, the different requirements for the line and the electricity infrastructure in that area, as appropriate. There is no cost associated with requesting a review of the consent but clearly a thorough understanding of the issues that would be considered should be reached in order to provide a coherent argument. Requesting a review prematurely may result in the Secretary of State determining that there were no grounds for a further review. The Secretary of State can only consider an application for review on the merits of the actual case presented. The development and characteristics of the area, the environmental impact of the line, and any alternatives, the existing electricity infrastructure and any proposed changes in that locality would all need to be fully assessed before requesting a review.

12.7 Conclusion

It will be seen that where a proposal is made for a new overhead line then the permissions required of the landowners and occupiers affected will need to be considered from different angles. Will there need to be a necessary wayleave hearing, as discussed in Chapter 10? Are alternative rights such as an easement or compulsory purchase as discussed in Chapters 15 and 11 respectively a better option? Are they the only option? Is there a need for planning permission and ministerial consent as discussed in this chapter carrying, as such solutions do, the potential for a public inquiry as discussed in Chapter 13? The matrix of all these options will need to be considered. Owners/occupiers, whether individual or corporate, need to have their views taken into account whether they are granting rights voluntarily, as discussed in Chapter 5, or objecting to a planning application. However unlikely it may seem, it is conceivable that such a landowner/occupier could agree terms for the electric line and yet object directly to the Secretary of State or the relevant local planning authority. Where terms were not agreed they could voice objections at a necessary wayleave hearing or under compulsory purchase proceedings as well as at a public inquiry considering the planning application for the line.

Public Inquiries

13.1 Outline of chapter

This chapter concerns public inquiries held in relation to applications for consent for overhead electric lines under section 37 of the Act by licence-holders in circumstances where objections have been made by the local planning authority. It will also address public inquiries convened under section 62 of the Act where objections have been made by persons other than the local planning authority and the procedure for reviews of existing consents.

13.2 Public inquiry distinguished from necessary wayleave hearing

This procedure for a public inquiry needs to be distinguished from the procedure applicable to a hearing held in relation to an application to the Secretary of State for a necessary wayleave as discussed in Chapters 8, 9 and 10. There will however be occasions where there will be a link between the two. The Secretary of State cannot grant consent under section 37 of the Act until all wayleaves have been obtained and, where not granted voluntarily, then granted by the Secretary of State as discussed in Chapter 9. As the procedural rules for the two applications are different, and may involve different parties, it would not be appropriate for the two hearings to be held together although they may be held contemporaneously. It is however open to the Secretary of State to issue a decision letter following an application under section 37 of the Act based on a decision he is "minded to" make. This would be particularly appropriate where a landowner owns a significant area affected by a proposed overhead line and has not granted a wayleave and the decision letter, if in favour of the application, would be conditional on the necessary wayleaves being obtained.

As discussed in Chapter 12 before a distribution or transmission licence-holder makes an application to the Secretary of State, it is required to consult with the local planning authority for the area in which the line is to be constructed. Where the local planning authority determines that the application should be refused, or suggests conditions to which the licence-holder objects, a public inquiry will need to be held. It is also at the discretion of the Secretary of State to hold an inquiry where qualifying objectors[1] object to the application.

The issues to consider at the inquiry may be wide ranging in nature but require an appointed inspector to hear the case on behalf of the Secretary of State and must only have regard to relevant factors. For the licence-holder these issues would relate to its obligations under its distribution or transmission licence to provide an efficient, economic and co-ordinated supply of electricity and also to its environmental obligations under Schedule 9 to the Act. These obligations require a licence-holder to have regard to the preservation of flora and fauna and additionally to mitigate the effect of its proposals on the countryside.

13.3 Procedure required

An inquiry will be called where the Secretary of State receives an application under section 37 of the Act to which the local planning authority objects, or in respect of which, either it seeks to impose conditions unacceptable to the licence-holder or an objection is lodged by a qualifying objector.

An application for consent under section 37 of the Act shall be in writing and shall describe by reference to a map the land to which the application relates, that is, the land across which the electric line is proposed to be installed or kept installed. The application shall also state the length of the proposed line and its nominal voltage and whether all necessary wayleaves have been agreed with owners and occupiers of land proposed to be crossed by the line and shall be supplemented, if the Secretary of State so directs, by such additional information as may be specified in the direction.

---

1   Rule 2 SI 1990 No 528 — The Electricity Stations and Overhead Lines (Inquiries Procedure) Rules 1990.

13.4 Environmental issues

Where the voltage of the proposed overhead line will be 275,000 volts or more the licence-holder needs to provide with the application an environmental statement in accordance with EC Directive number 85/337/EEC as implemented by The Electricity and Pipe-line Works (Assessment of Environmental Effect) Regulations 1990,[2] or as specifically directed by the Secretary of State under Rule 3(2) and 5(1) of these regulations.

13.5 Pre-inquiry procedure

Under the Electricity Generating Stations and Overhead Lines (Inquiries Procedure) Rules 1990,[3] notice must be given of the inquiry in the local press. A suitable venue must be selected in the locality having allowance for the facilities required for the Inspector, who may also sit with an inspector appointed from another government department such as the Department for Communities and Local Government, where planning issues also need to be addressed. There will also need to be a provision for a transcript writer and recording facilities. The buildings typically chosen are public buildings with the ability to close off the room from other distractions and provide sufficient space to allow representation by parties who have lodged objections, space for the representatives of the licence-holder and for members of the general public. This requirement can be distinguished from the opportunity given by the Electricity (Compulsory Wayleave) (Hearing Procedure) Rules 1967,[4] where the hearing may be held in private.

13.6 Pre-inquiry meeting

Where the Secretary of State considers it desirable to hold a pre-inquiry meeting then this will be convened.[5] The licence-holder, local

---

2 SI 1990 No 442 as amended by the Electricity and Pipe-line Works (Assessment of Environmental Effect) (Amendment) Regulations 1994.
3 SI 1990 No 528 as amended by the Electricity Generating Stations and Overhead Lines and Pipe-Lines (Inquiries Procedure) (Amendment) Rules 1997 [SI 1997 H2].
4 See Chapter 10.
5 Rule 5, The Electricity Generating Stations and Overhead Lines (Inquiries Procedure) Rules 1990 SI 1990 No 528.

planning authority and any other qualifying objector will be served with a notice indicating the intention to cause a pre-inquiry meeting to be held together with a statement of the matters to be considered. "Qualifying objector" means the local planning authority and any person who has made a written objection to the application to the Secretary of State prior to the Secretary of State's written notice of his intention to cause a public inquiry to be held.[6] A licence-holder is required to publish in a newspaper local to the land in question stating the Secretary of State's intention to cause the pre-inquiry meeting to be held and a statement of the relevant matters including any issues raised by any other government department in relation to the application. No later than eight weeks after the Secretary of State's notice of the pre-inquiry meeting the relevant planning authority and the licence-holder are required to serve an outline statement on each other and also on the Secretary of State containing the issues they consider relevant. There is also the ability for the Secretary of State to require any other qualifying objector also to serve an outline statement of their case on the Secretary of State, the licence-holder and the relevant planning authority within four weeks. The pre-inquiry meeting is to be held no later than 16 weeks after the Secretary of State's decision. Twenty one days written notice is to be given of the pre-inquiry meeting to the planning authority, licence-holder and any qualifying objector and a wider remit is provided to the Secretary of State to include any other person whose presence at the pre-inquiry meeting would seem, to the Secretary of State, to be desirable.

13.7 Role of inspector

The inspectors are appointed by the Secretary of State from within the Engineering Inspectorate of the Department of Trade and Industry and all are experienced electrical engineers and have been given training in the relevant procedural rules and in controlling hearings and inquiries. The appointed inspector is required to act in accordance with the principles of fairness, openness and impartiality and is responsible for ensuring that the hearing is run efficiently and time used to the best advantage of all concerned. He must prepare, conduct and subsequently report on the proceedings with dispassionate

---

6   Rule 2, The Electricity Generating Stations and Overhead Lines (Inquiries Procedure) Rules 1990 SI 1990 No 528.

moderation. He has the ability to intervene where he considers it helpful, particularly to clarify any issues raised. It would be normal practice for the inspector to examine witnesses after the licence-holder and objectors have finished their questioning.

Prior to opening the main inquiry the inspector will require confirmation that the statutory requirements relating to advertising the inquiry have been complied with and the licence-holder has published a notice stating the place, date and time of the inquiry in one or more widely read newspapers circulating in the locality in which the land is situated, on two successive weeks,[7] and where appropriate, an environmental statement has been provided as discussed in chapter 12. The inspector will also confirm that the timescales have been complied with in giving notice of the inquiry as provided by The Electricity Generating Stations and Overhead Lines (Inquiries Procedures) Rule 1990[8] as amended by The Electricity Generating Stations and Overhead Lines and Pipe-lines (Inquiries Procedures) (Amendment) Rules 1997.[9] Confirmation is also required that any objections to the application were received within the correct period of time. The time period is, if the nominal voltage of the proposed overhead line is not less than 132,000 volts, within two months of the notice of application from the relevant planning authority[10] or within 28 days by other persons.[11]

In the case of applications under section 37 of the Act the requirement is that a statement of case is prepared by the licence-holder and the relevant planning authority on each other and to the Secretary of State and any other qualifying objector no later than six weeks after the relevant date provided by the Secretary of State. Where an inquiry is held under section 62 a statement of case is not required.

Documents referred to in the statement are also required to be produced so that each party is provided with all relevant information. The inspector will also confirm that all parties have been informed of the date, time and place of the inquiry and given the correct period of notice — 28 days notice in writing in accordance with rule 10(3).[12] Where a case is particularly extensive and a written summary of

---

7   Paragraph 4(1)(c), Schedule 8, Electricity Act 1989.
8   SI 1990 No 528.
9   SI 1997 No 712.
10 Rule 8(1) SI 1990 No 455.
11 Rule 7 SI 1990 No 455.
12 SI 1990 No 528.

statements of evidence has been requested in order to facilitate the better conduct of the inquiry in accordance with Rule 13 (1)[13] then the inspector will also verify that the written summary has been copied to all other parties.

The inspector, having set out at the pre-inquiry hearing the times of opening and closing of each day's sitting and with breaks where required, will also reaffirm the procedure for the giving and examining of evidence and for opening and closing submissions.

13.8 Role of advocates

Where the issue requires it, it is preferable for an advocate to present the case of the licence-holder, the planning authority and any other interested parties in order to present a coherent and cohesive argument. This does not preclude individuals from presenting their own submissions but the experience of a good advocate in presenting relevant evidence in a structured manner will be of considerable assistance to the inspector in compiling his report and arriving at a balanced recommendation. An advocate will also be experienced in cross examining the evidence submitted by witnesses and sifting out any irrelevant aspects.

13.9 Typical flow of events

At the opening of the inquiry the inspector will confirm the identity of those present, review any arrangement agreed at the pre-inquiry, check if statements of case have been exchanged, where that has been agreed, and then ask the licence-holder's representative to open his case. This opening submission will set out (in general terms) the proposals of the licence-holder and provide a skeleton argument for its proposals. It will also reassert who is to provide evidence for licence-holder and on which aspects. Following that the first witness of the licence-holder will be asked to present his evidence. This is usually in the form of reading out a written statement prepared beforehand which would already have been provided to all objectors and relevant parties. The advocate will give the witness the opportunity to present any further evidence or clarify the written evidence provided before any objector is allowed to

---

13 SI 1990 No 528.

cross examine the evidence of the witness. Following cross examination the opportunity exists for the licence-holder's advocate to re-examine the witness in order to deal with points of clarification for the benefit of the inspector.

13.10 Closing statement by objectors

After the presentation of evidence, examination and cross-examination the opportunity is given for all objectors, including the planning authority, to make their closing submissions. This is to provide a conclusion to their arguments and, without raising new evidence, attempt to summarise the main issues they have put forward.

13.11 Closing statement by the licence-holder

The advocate for the licence-holder will summarise the case they have presented, clarify the points made and comment on the strengths or otherwise of the objector's case and conclude their proposal.

13.12 Closing of inquiry

Following the final submission for the licence-holder the inspector will provide a final opportunity for any clarification, set out his role after the inquiry and then formally close the public inquiry. It is normal practice for an inspection of the site to be made and if attended by either party the inspector would give all objectors an opportunity to be present. A qualifying objector can only accompany the inspector's site visit if they attended the inquiry.[14] The inspector will also clarify that no evidence can be made at the site visit as the inquiry has formally concluded. The inspector is however likely to ask for clarification on any factual point from the site visit. As a matter of course it is likely that the inspector will have already inspected the site in question prior to the inquiry as provided for in Rule 15.[15]

---

14 Rule 15(3) SI 1990 No 528.
15 SI 1990 No 528.

13.13 Remit for objections

The practical difficulty for objectors in these cases is that their primary objection may relate to a range of disparate issues, including the perceived health effects, interference with television reception or financial loss rather than the application itself. While an inspector gives a wide remit for objectors to make their point, in making a recommendation to the Secretary the State only matters relevant to the application can be taken into account. This procedure could in theory be used by a landowner who has failed to grant a voluntary wayleave and who also objects at the public inquiry to the line. In such an eventuality a necessary wayleave would not be granted by the Secretary of State until the application under section 37 of the Act had been determined. It is therefore possible for a decision on both the necessary wayleave and the section 37 application to be entertained at the same time as clearly to make a determination on one aspect only by the Secretary of State would prejudice the decision on the outstanding application.

13.14 Recommendation to the Secretary of State

Upon conclusion of the inquiry the inspector will prepare a written summary of the investigation having reviewed the transcript of the inquiry and taken evidence and representations into account. The recommendation will say whether the application should be approved, and if so, whether with or without modifications. The Secretary of State is not obliged to follow the inspector's recommendations but in disagreeing with the inspector's recommendation he must make clear the reason for doing so.

13.15 Consent review

Where consent is granted under section 37 of the Electricity Act 1989 or Schedule 10b of the Electric Lighting (Clauses) Act 1899 then provision is made for the consent to be reviewed after five years of being granted.

13.16 Inquiry costs

Normally for most tribunal and inquiry hearings the costs of the government inspector are prescribed by statutory instrument. This covers work in relation to inquiries at the instigation of local authorities, highway agencies or the Environment Agency and the provision for fees are covered in the Fees for Inquires (Standard Daily Amount) Regulations 1990.[16] However, the regulations do not extend to work under the Act and therefore the costs of the inspector are the responsibility of the government department concerned; in this case the Department of Trade and Industry. The licence-holder and any objectors must each bear their own costs and there is no provision for claiming costs from other parties.

---

16   SI 1990 No 2027.

Part 4

Financial Aspects

Wayleave Payments

14.1 Outline of chapter

This chapter explains:

- why wayleave payments are made
- who the beneficiaries of the wayleave payments are
- the calculation of payment values
- the basis of making payments under wayleaves.

14.2 Introductory overview

The responsibility for making wayleave payments lies with the licence-holder whose equipment has been installed and retained on private land. There needs to be a contract in place requiring the licence-holder to make payments to the landowner and/or occupier. The contract can be found in one of three forms: under the terms of an express non-statutory wayleave, in the form of a necessary wayleave granted by the Secretary of State or, where an existing wayleave has been adopted by a new landowner/occupier of land affected by electric lines, in the form of an implied non-statutory wayleave.

Express non-statutory wayleaves, as defined in Chapter 6, granted by landowners and/or occupiers, ("grantors"), provide for payments to be made in return for the electricity apparatus on their land. Such wayleaves also set out what the payment will be, how it will be calculated and the period covered by the wayleave payment. Under the terms of the vast majority of express non-statutory wayleaves, the payments are made annually and, where a grantor has numerous wayleaves with the licence-holder, then normally all payments will be made at the same time. A grantor who has entered into an express non-statutory wayleave therefore expects to be paid the rates, either at the amount set by the wayleave, or at an amount set following a review provided for within the contract.

A review of the payment level applies in the vast majority of cases.[1] A national scale of wayleave payments was established in the 1960s identifying a payment to the owner and a separate payment to the occupier. These payments are largely based on agricultural values. Certainly the occupier payments were the result of a detailed analysis investigated by the Agricultural Development and Advisory Service (ADAS) on the costs associated with working around overhead line supports, which produced a national average.

Payments under necessary wayleaves granted by the Secretary of State are distinct from an agreement between the landowner/occupier and a licence-holder as they need not be an annual payment, as provided by most voluntary wayleaves, but can be assessed on a periodic or lump sum basis for the term of the necessary wayleave.

14.3 Calculation of wayleave payments

The quantum of the wayleave payment is based on the type and size of the structure permitted by the non-statutory wayleave.[2] Payments had previously existed for wood pole lines based on rounded units of currency typically of one shilling (5p) for a pole and 6 old pence (2 and a half pence) for each stay wire. These payments were generally for lines of lower voltage over agricultural land in the early 20th century. Scales of payment to apply on a national basis appear to have commenced in 1933 following discussions between the Central Electricity Board, (forerunner to the Central Electricity Generating Board) now the National Grid. The only distinction drawn was between rental (representing payments to landowners), and compensation (to reflect the interference to agricultural operations). This latter element further split down between arable and cultivated grassland. The distinction between the two was a multiplier of approximately four times the

---

1 Please refer to Chapter 6, section 6.5.
2 Chapter 3 discusses these different structures in detail. The relevant fee scales are set out in Appendices 19 and 20.

cultivated grass land rate for arable land. The typical size of towers at that time related to the 132,000 volts system and the largest tower size identified was 20 ft sq. In the 1950s larger tower sizes were introduced for the 275,000 volts overhead lines and subsequently, in the 1960s, even larger tower sizes were introduced to cater for 400,000 volts overhead lines. Also during the 1960s national payment levels were fixed for lower voltage line apparatus. All of the rates reflected agricultural land values as it was no doubt seen that the vast majority of lines crossed agricultural land. Until the 1960s these figures were considered relatively nominal and the perceived concern of landowners and occupiers related more to the actual location of the equipment rather than financial recompense. However, in the 1960s, following rural electrification in the preceding decade and the significant growth in electricity demand, further attention was given to these payment levels.

In 1966 a joint investigation was undertaken by the Central Electricity Generating Board in conjunction with the Area Electricity Boards (responsible for lower voltage lines) and with the National Farmers Union and Country Landowners Association. In 1982 there was a challenge to the rental levels and they were considered at length in the case of the *Trustees of the Clouds Estate* v *Southern Electricity Board*.[3] This case was determined at the Lands Tribunal following the claim by the Clouds Estate that their land merited the highest land value in England and Wales and therefore the rental level should be higher to reflect that. This created the need for a further increase in payment to landowners. A further investigation was subsequently considered appropriate, largely due to the high level of inflation at that time together with the increased costs of agricultural operations. A preliminary investigation was undertaken and produced an increase, largely in compensation payments. In 1994 under challenge from the National Farmers Union and Country Landowners Association (now Country Land and Business Owners Association) the licence-holders commissioned a comprehensive study to be undertaken by ADAS, (at that time part of the Ministry of Agricultural Fisheries and Food, now an independent organisation). A detailed study was undertaken of a variety of agricultural operations reflecting the significantly increased size of machinery then in operation compared with the 1950s and 1960s.

---

3   LT 183/1982.

An average compensation figure was applied which was inevitably distorted on a regional basis in comparing the large open arable areas in East Anglia with small dairy farmers in North Wales, which is clearly anomalous. However, the consensus view prevailed in applying this average with provision for particular situations to be considered on their individual merit. Further distinctions were provided in respect of structures on boundaries interfering with hedge cutting operations and for special situations where extensive grassland operations were undertaken and in orchards and hop gardens. A further distinction was provided where two overhead lines existed in the same field creating increased difficulties and at that time it was agreed that for fields of less than 20 acres each line would have the rates enhanced by 50% and, if a third line, a doubling of payment for each overhead line support. Subsequently, this guideline has changed to an enhancement of all payments by 50% where supports of separate lines in the same field are within 30 m of each other.

14.4 Underground cable rates

As most of the attention was placed on the interference with agricultural operations, little regard was given to the extent of underground cable payments as these rarely applied in rural situations. A nominal payment of 5p was generally applied but this often had no regard for the length of line or particular location. While it may seem anomalous, again agricultural values were applied in the mid 1990s and averaged in order to arrive at a formula assessment for underground cable payment values. Sections of land affected by cables were applied in 50 m lengths and the area affected was considered to be a width of 2 m; effectively 1 m each side of the cable, with a nominal provision for the need for access to a further 1 m each side of that. As the rate for permanent easements for underground cables was generally taken to be two thirds of the freehold value for an annual wayleave, a rate of one third of land value was taken. Applied to that was the annual equivalent of 5%; a multiplier commonly used by licence-holders in converting annual payments to permanent easement rates. In compiling these assessments in 1996 the rate applied to a 50 m length on an annual wayleave basis was £1.15p where the average land value per hectare applied was £6,000.

Cable rates for higher voltage lines, 400,000 volts and 275,000 volts, are more likely to be on a permanent easement basis, largely due to the high cost of relocation where a payment related to the freehold value should be applied.

Given the extensive nature of the ADAS study it was considered uneconomic to revise this on a regular basis and therefore an annual review was undertaken applying increases in agricultural operations taken from the *John Nix Farmer Management Pocket Book* and what is now the DEFRA census.

14.5 Basis of payment to owner

The term rent has been applied to the annual payment to landowners, and is a generic term used throughout the electricity industry. Strictly speaking the payment under a wayleave is the compensation to the landowner or occupier for granting the non-statutory wayleave. Rent is more commonly applied to exclusive possession of demised land or premises. It can be seen that the area occupied by a pole, approximately 1 ft diameter, cannot be used for any other purpose, but clearly the restrictions caused by the conductors which oversail adjacent land also need to be reflected. However, as the term has been in use and applied to wayleave payments, it has become accepted parlance. Additionally under UK taxation law it is accepted as a rental payment and was therefore subject to the rent freeze in 1974 applied by the government and continues to be subject to income tax as with other property rents. The continuation of payment at a rate different from that originally agreed can convert an express contract into an implied wayleave in the circumstances described in Chapter 18. Failure to continue this payment can be seen as not only a breach of the contract but may also prevent the continuation of the agreement. The significance of this is considered further in Chapters 6, 7 and 8.

Further information was provided on the calculation of payment rates by the Country Landowners Association in September 1997 in their handbook entitled *Cables and Wires*. The basis behind the payment levels was considered and these were further extended and utilised by the Post Office, now British Telecom, for their apparatus which has a similar effect on agricultural operations. Further evidence of the extent of these rates, effectively giving them national acceptability, was applied in the expansion of telecommunications apparatus during the 1990s and separate payments were established for fibre optic cables strung on existing overhead electric lines. The rates were applied on a per structure basis as this was a more useable reference point for payment rates and the identification of assets which was more reconcilable than with the length of telecommunication cable across the landowner's property. In comparison to electricity wayleave payments, relatively generous payment levels were offered where fibre optic cables could be attached to electric lines. The rights for fibre optic cables are covered, when for third party use, under the Telecommunications Act 1984. Where "telecommunication" cables are strung or installed by licence-holders, then the Electricity Act 1989 at section 64, includes these cables within the definition of electric line under that Act, but only when used solely for electricity operational purposes by the licence-holders themselves. They cannot therefore be assessed in compensation terms under the Telecommunications Act 1984.

14.6 Overhead lines through woodland

Due to the specific nature of forestry operations a separate rate was established for electricity lines through woodland and forestry with rates agreed in the past with the Forestry Commission. These generally reflect the inability to plant and the need to keep areas clear underneath overhead lines within the safe working clearance. These are therefore expressed in terms of the area of land used on a rate per hectare rather than in terms of single items of apparatus.

14.7 Payments following necessary wayleave granted by Secretary of State

As described in Chapter 9 where the Secretary of State grants a necessary wayleave then it is left for the landowner and occupier to agree with the licence-holder the amount of compensation; either as a lump sum or periodic payments. The date the Secretary of State grants the necessary wayleave provides the date of valuation and interest will be payable on the amount assessed in accordance with the Acquisition of Land (Rate of Interest After Entry) Regulations 1995.[4] This provides for simple interest to be added from the grant of the necessary wayleave to the date full payment is made. It follows that if a partial assessment is made then simple interest will be added to that from the

---

4   SI 1995 No 2262.

valuation date to the date of payment and on the balance until the settlement of compensation on the whole. It needs to be borne in mind that where a lump sum payment is made, it is on the basis that it would be for a fixed period, currently 15 years, after which the necessary wayleave granted by the Secretary of State can be reviewed. While the line may well remain, if notice is given after the period of review, there is no contract in place to provide for a continuing payment.

This is covered under paragraph 7, schedule 4 to the Act which provides that where a wayleave is granted to a licence-holder both the occupier of the land and the owner may recover from the licence-holder compensation in respect of the grant. Additionally a claim for damages may be recovered as provided by that section as follows:

> Where in the exercise of any right conferred by such a wayleave any damage is caused to land or to moveables, any person interested in the land or moveables may recover from the licence-holder compensation in respect of that damage; and where in consequence of the exercise of such a right a person is disturbed in his enjoyment of any land or moveables he may recover from the licence-holder compensation in respect of that disturbance.

Applications to the Lands Tribunal are discussed in Chapter 16. Moveables are defined as chattels[5], that is to say an item of personal property distinct from land and property.

Any dispute as to compensation is covered by clause (4) of paragraph 7 and falls to be determined by the Lands Tribunal in accordance with sections 2 and 4 of the Land Compensation Act 1961.

Where payments are made under a voluntary wayleave, provision for payment is provided within the contract and where that contract is brought to an end then there is no continuing provision for payment. This creates an anomalous situation. The existence of the electric line imposes a continuing impact on a landowner yet there is no provision under legislation, in those circumstances, for compensation to be paid. While it is the policy of licence-holders to continue to offer the previous rates of payment where wayleaves have been terminated, there is no statutory obligation for them to do so. What is required is provision for compensation payments over the period of time from the termination of a voluntary wayleave to the grant of a necessary wayleave by the Secretary of State. This is discussed further in Chapter 18.

---

5   Paragraph 12 of Schedule 4 to the Electricity Act 1989.

14.8 Wayleaves compensation scales

The following is an extract from a press release issued on behalf of all licence-holders. It details the level of payments made in different agricultural situations and provides for the review of the owner's payment. The values are heavily biased towards agricultural land. New wayleave compensation scales have been recommended to compensate landowners for the presence of licence-holders' apparatus on their land.

The licence-holders have concluded discussions with the Country Land & Business Association, the National Farmers Union and the Farmers Union of Wales and the schedules of new rates are set out (at Appendix 18 and 19).

Wayleave compensation comprises two elements divided between landowners and occupiers — owner/occupiers receive both payments.

The recommendation as to the revised scales payable to landowners will take effect from 1 April 2005 for a four year period to 31 March 2009. Farmers do not need to apply to their electricity company for the increases.

The recommendation as to the occupiers' payment scales will continue at existing levels pending the outcome of continuing research by ADAS and discussions between licence-holders and the Country Land & Business Association, the National Farmers Union and the Farmers Union of Wales. Should this result in a recommended increase in payment, the recommendation will be the subject of a future press release.

The lines only payment is to apply to any length of line where no supports are located upon the grantors land.

Rates for PB (Painter Brothers design) single-unstayed are the same as poles. A PB double-unstayed assumes a double limb structure with a gap between the two limbs up to and including 6 metres and a PB multi-stayed assumes up to and including 12 stays based on a ground area of 23 metres square.

Payments for underground cables will be based on a minimum length of 50 metres or part thereof. Associated pilot, control or earth cables are included in this rate. Allowance has been made for the average area where caution should be exercised in carrying out agricultural operations.
Where supports of two separate electricity lines are within 30 metres of each other in the same enclosure the occupier compensation payment for those supports will be increased by 50%. This only applies to supports for which annual wayleave payments are made.

Enhanced payments for Commercial orchards and hop gardens where supports interfere with the movement of machinery are to be paid as follows:–

Hop Gardens	—	150% arable rate
Commercial Apple and Pear	—	150% arable rate
Commercial Cherry, Cider and Plum	—	150% permanent pasture rate

Farming systems which include crops involving exceptionally intense and multiple cultivations, field irrigation or other special factors may be considered for enhancement of standard rates on merit. Any such claims should be sent to the relevant licence-holder.

Easements

15.1 Outline of chapter

In Chapter 5, we outlined an alternative to a landowner or occupier granting a non-statutory express wayleave, or (the mirror image) to a licence-holder pursuing the grant of a necessary wayleave; namely the agreement of terms for an easement by deed of grant. Wayleaves are the method generally utilised by licence-holders to install and keep installed electric lines on private land but there are situations where a wayleave cannot be obtained. We considered those circumstances and the benefits to both landowner or occupier and licence-holder respectively of an easement as opposed to a wayleave and explained that if agreement is reached as to the terms of the grant, a joint reference can be made to the Lands Tribunal for the determination of compensation. This chapter gives further consideration to that alternative solution against the following:

- the legislative background
- the various methods of assessment of compensation (which are as applicable to statutory arrangements as they are to voluntary ones)
- the factors to take into account in arriving at the consideration.

15.2 Legislative background

Most compensation payments are agreed voluntarily with the licence-holder in the shadow of compulsion. The statutory rights are contained in schedules 3 and 4 to the Act. Schedule 3 to the Act also deals with the compulsory acquisition of land for a licence-holder's operational purposes. In situations comparable to the licence-holder's need for wayleaves, the assessment of compensation takes account of the three heads of claim normally encountered in compulsory acquisition, namely land taken, injurious affection and disturbance. Such assessment embraces the compensation code contained in the six rules within section 5 of the Land Compensation Act 1961. It is not the intention of this book to give a full explanation of compulsory purchase legislation which is more comprehensively covered in other dedicated works on the topic.[1] However to assist the reader a few explanations in laymen's terms may help. Land taken refers to the land expropriated from the original owner and in relation to the work of licence-holders is explained more fully below. Injurious affection may be expressed as the damage to the interest in the retained land of the owner caused by the use and presence of the licence-holder's apparatus, including the owner's loss of enjoyment of the land. Disturbance relates to the physical damage caused by installing and maintaining the apparatus. The Court of Appeal has held[2] that a claim for loss of profits is not too remote in law to be included in such a compensation claim.

Where it is not possible to agree the quantum of compensation, then a reference is made to the Lands Tribunal under the Lands Tribunal Rules 1996.[3] Such references are discussed in Chapter 16. It is also possible where most terms are agreed, but crucially not the consideration, for a reference to be made by consent to the Lands Tribunal to determine the amount of compensation. Such references are comparable to the determination of rights for a compulsory purchase order for an electric line over land where the terms are set down in a compulsory purchase order but the amount of compensation has not been agreed.

As already stated, the assessment of compensation for an easement taken under Schedule 3 to the Act follows the statutory principles. The same principles apply to assessing the extent of the land taken, the injurious affection and the disturbance. Where it is not possible to reach agreement then a reference can be made to the Lands Tribunal.[4]

---

1   Such as Barry Denyer-Green *Compulsory Purchase and Compensation* (8th ed).
2   *Welford* v *EDF Energy Network (LPN) Ltd* [2007] EWCA Civ 293.
3   SI 1996 No 1022.
4   This is explained in Chapter 16.

15.3 The compensation code

Where work is carried out by a licence-holder using its own statutory powers, the assessment of compensation has to take account of the statutory background. These statutory rules,[5] referred to as the compensation code, aim to help determine open market value. The aim is to avoid a licence-holder being held to ransom or a landowner enjoying an enhancement in value due solely to the work being taken into account. The rules are set out below with necessary adaptations.

(1) No allowance shall be made on account of the acquisition being compulsory.
(2) The value of land shall, subject to the further provisions below, be taken to be the amount which the land if sold in the open market by a willing seller might be expected to realise.
(3) The special suitability or adaptability of the land for any purpose shall not be taken into account if that purpose is a purpose to which it could be applied only in pursuance of statutory powers, or for which there is no market apart from the requirements of any [licence-holder] possessing compulsory purchase powers.
(4) Where the value of the land is increased by reason of the use thereof or of any premises thereon in a manner which could be restrained by any court, or is contrary to law, or is detrimental to the health of the occupants of the premises or to the public health, the amount of that increase shall not be taken into account.
(5) Where land is, and but for the compulsory acquisition would continue to be, devoted to a purpose of such a nature that there is no general demand or market for land for that purpose, the compensation may, if the Lands Tribunal is satisfied that reinstatement in some other place is bona fide intended, be assessed on the basis of the reasonable cost of equivalent reinstatement.
(6) The provisions of rule (2) shall not affect the assessment of compensation for disturbance or any other matter not directly based on the value of land.

15.4 Line installation date

The assessment of compensation for existing electric lines is further complicated by the need to know the date the electric line was built.

---

5 Section 5 of the Land Compensation Act 1961.

This can have an impact on the assessment of injurious affection to the land caused by the electric line. This may need more research than simply confirming the date of the original construction of the line. It is possible that parts of the electric line, such as the conductors, may have been changed or the route amended. In those circumstances, the electric line would have needed a further consent under section 37 of the Act. In this connection we refer the reader to Chapter 12. The date of any such consent would be the relevant date for determining "the date of construction" In some cases, therefore, it could be demonstrated that the electric line was constructed post 17 October 1972 and therefore the effect of the entire line, in compulsory purchase terms, could be taken into account.

15.5 Section 44 of the Land Compensation Act 1973

Where licence-holders are acting as acquiring authorities whether in a voluntary transaction under the shadow of compulsion or in the formal exercise of their rights under the Act, then as part of having regard to the compensation code,[6] they must also take account of the Land Compensation Act 1973. Under this Act an anomaly arises. The imposition of section 44 of the 1973 Act has the consequence that for electric lines constructed prior to 17 October 1972 the loss to the landowner can only be assessed in respect of that part of the line actually crossing the landowner's property. Clearly lines cover long distances and those unaware of section 44 would naturally assume that the overall impact of the line would be taken into account.

By way of example if the owner of Whiteacre has by virtue of a non-statutory wayleave an overhead line crossing his land which has been constructed before 17 October 1972 and then grants an easement to the licence-holder for the line, there is a limitation on the assessment of injurious affection. If the overhead line crosses Blackacre, which is adjacent to Whiteacre and the fact that the line crosses Blackacre reduces the value of Whiteacre due to the visual impact on Whiteacre of the overhead line across Blackacre, then that reduction in value has to be taken into account.

Let us assume that Whiteacre has a value of £2,000,000 if no overhead line at all were present. Let us then assume that the injurious

---

6   As explained in the preceding section.

affection caused to Whiteacre by the overhead line across Blackacre reduces the value of Whiteacre to £1,800,000. A valuer would need to dismiss the fact that the electric line continues and likewise would need to ignore the impact of the tower situated on the adjacent land. In assessing the reduction in the value of Whiteacre caused by the electric line within the boundaries of Whiteacre which the easement permits, the starting point is £1,800,000 not £2,000,000. There are numerous ways of assessing the reduction in value caused by the easement. Suppose for the purpose of this example it were demonstrated that the section of the electric line across Whiteacre reduced the value of Whiteacre by 10%, then the assessment of compensation for injurious affection would be £180,000 (£1,800,000 × 10%) and not £200,000. This is because section 44 of the Land Compensation Act 1973 disallows the full reduction in value if the electric line (or the "works" to use the term from that Act) were constructed prior to 1972. If neighbouring towers on adjacent land are not to be taken into account then one can envisage the hypothetical situation in which the valuer has to consider wires in the air unsupported by any tower! For electric lines constructed after 17 October 1972, the Land Compensation Act 1973 does allow the full extent of the works to be taken into account. Whether the legislation should have been retrospective is something we consider further in Chapter 22.

The objective of section 44 of the 1973 Act was to overrule the decision in *Edwards* v *Ministry of Transport*,[7] where it was determined that land taken for a highway scheme had to be assessed on the use of the land for the purpose of the underlying scheme. In that case the land taken was to be utilised as a grass verge for the highway. Therefore the injurious affection to the dispossessed owner had to be assessed on the basis of the impact the grass verge would make on the retained land not the impact of the new highway in its entirety. This was clearly illogical and therefore the 1973 Act was introduced in order to allow the full impact of the scheme to be taken into account in arriving at the true measure of loss in terms of injurious affection on the retained land of the dispossessed owner. Where the works are a new line yet to be constructed then clearly this produces an equitable basis for the assessment of compensation. However when new rights are acquired for existing works the legislation produces this perverse result.

---

7   *Edwards* v *Minister of Transport* [1964] 2 QB 134.

15.6 Compensation payments for easements

As indicated, an alternative to a voluntary wayleave being entered into by a landowner or the licence-holder being awarded a necessary wayleave, an easement by deed of grant can be entered into.

This can either be on the basis of a grant in perpetuity or for a term of years. In both cases the rights are registered and bind successors in title and therefore provide security for the licence-holder in retaining that particular electric line. With an easement for a term of years the valuation will need to reflect the loss over the term granted and will also need to include a basis for renewal or removal of the electric line in specified circumstances.

15.7 Principles of valuation

Where landowners seek to capitalise a wayleave agreement, and while it is possible for a lump sum payment to be made, it would be more appropriate for a permanent easement to be entered into. This then allows the six rules of compensation[8] to be applied as the principles to be followed are included in Schedule 3 to the Act. Where a permanent easement either is not agreed or cannot be granted, for example in cases where only a long leasehold interest is held, then a term easement can be entered into up to the period of time provided for within the lease.

15.8 Land taken

The area occupied by a tower is generally seen as being entirely taken by the licence-holder and therefore the dimensions of the tower should be measured at the ground level of the tower foundations. As the foundations are likely to extend outwards from the legs of the tower, it may be possible to show that a greater area needs to be taken into account to allow for the support of the ground around the tower base. In compulsory purchase situations licence-holders have applied the area required for the frustum of support for the tower to be included and this has extended the area considerably. Notionally a further area of 3 to 5 m from the base of the tower legs should be added to each side

---

8   Section 5 of the Land Compensation Act 1961 and section 15.5 above.

of the tower. Where the tower is particularly tall, or the angle in the line is significant, larger foundations are likely to have been installed and an even larger area may need to be calculated. With an existing tower it would be reasonable to request the details of the tower foundations from the licence-holder to ensure that the full area occupied, and therefore the extent of land effectively taken, is properly considered.

15.9 Injurious affection

The greatest imposition of electric lines is in the restriction to development in proximity to the conductors. This applies to the land immediately beneath the conductors and the land beneath the outward swing of the conductors referred to as their sag and swing. This is the area affected by the conductors being greater than the physical area perceived when in still air. An allowance needs to be made for the movement of the conductors, which takes into account their physical expansion due to electrical load and prevailing weather conditions, which increase the length of the conductors, dependent on their thermal capacity, and then applying the potential for the line to swing to 45 degrees out from their vertical line on the insulators. This creates an elliptical effect along the route of the line where the conductors swing out at their greatest distance at mid span between supports. These conditions dictate the safety clearances applied by licence-holders and, for specific situations, details can be obtained from the licence-holder in order to assess the extent of physical development possible. This is also known as the swing envelope of the line so that the area above and around the conductors is taken into account, where appropriate.

There is also the visual effect on adjacent land owned by the same landowner to be taken into account. Where it can be demonstrated that specific development cannot be accommodated beneath the line, then the loss in land value as applicable to those development areas could be used as a basis of assessment measured against the alternative use that the land could be put to.

15.10 Disturbance

A number of items can be claimed under this heading which relate to the physical limitation of the land beyond simply land value. These will include the inability to erect metal fences running parallel with the line given the induced current possible, the inability to use certain plant beneath the conductors due to restrictions on safe working clearances, the need for safety equipment such as goal posts and warning signs and additional design costs in mitigating a developer's loss. All of these principal heads of claim should be accepted by licence-holders in cases where an easement is being granted. Additional consideration should also be given to the terms of the easement, particularly if a lift and shift[9] clause is included and the particular terms of that clause. The valuation of wayleave payments invariably relates to an annual loss and therefore one-off items are not easily reconciled. In addition all professional fees can be claimed, as applies with claims for loss in land value and injurious affection.

15.11 Development sites

Where an overhead line or underground cable restricts the full development potential of land, whether for residential, industrial or commercial purposes, then the normal assessment of compensation; including land taken, injurious affection and disturbance may not properly reflect the loss to the landowner or developer. In such situations it would be reasonable to produce a before and after valuation. This is comparable with the assessment of compensation for compulsory purchase where the rights are required by acquiring authorities, for example, in the case of highway and regeneration schemes. In highway schemes the entire interest of the affected land is acquired and the dispossessed owner can claim for the land taken and injurious affection to the adjacent land as well as any impact on severance of the retained land. With the linear rights obtained by licence-holders it is only at support positions where land can be said to be taken from the landowner and the intervening section beneath the conductors is limited in its subsequent use but the owner is not entirely dispossessed of the use. To assess the before effect of the line, it would be necessary to establish the extent of development for which planning consent could be obtained if the electric line were not present. This would need to take into account all relevant planning considerations and market conditions to produce a layout that could or would achieve planning consent if the electric line were not present. Again in comparison with other acquiring authorities, such as for highways and regeneration, the licence-holder's use, with an existing electric line, is

---

9   Explained in section 15.15 below.

already established. With highway schemes the acquisition of land takes place following confirmation of the compulsory purchase order and notice to treat having been served on the landowner.

As in other compulsory acquisition cases it is the responsibility of claimants to mitigate their loss. That would require owners to submit a timely claim and mitigate the effect of the line on their land by reasonable means. In view of that it would be inappropriate to propose high value development in proximity to towers where a greater loss would result if it were reasonable for the high value property to be elsewhere on the development. On the other hand, it would be appropriate, and a reasonable head of claim, for additional costs incurred by a landowner to be submitted as a head of claim. These would include relevant costs incurred in additional planning application fees, the need for further design and any input required by other consultants such as architects, planners, surveyors and lawyers.

15.12 Mineral extraction

Where minerals exist beneath the base of towers, and also under the land within the sag and swing of the conductors, then the presence of the line will diminish the ability to extract the minerals. This can extend to coal workings, lime pits and sand and gravel. The extent of loss can be calculated by identifying the area of land to be left undisturbed for each line support and then the quantification of the minerals lost. Allowance would need to be made for the cost of extraction and balanced against this is also the loss of the void that would be created which, where planning consent is granted, could be used for tipping or similar purposes. Beneath the conductors there will be restrictions on mechanical plant working under the line and heads of claims should be included to allow for double handling, additional design costs, modification to haul routes, the inability to extract from underneath the line in a safe manner and the need to provide warning signs and safety fencing. Where the after use of the site following extraction is proposed for uses alternative to the original, then consideration needs to be given to the quantification of future losses.

15.13 Tree planting and cutting

Within the terms of an easement there is also a restriction on planting and allowing trees to grow beneath an overhead line that would impact on the safety clearance. The clearance required tends to be more excessive than that applied to buildings due to the natural spread of trees which will invariably vary with the species. The allowance generally aimed for is the maximum likely height of the tree at its falling distance plus a safety tolerance. It is therefore not uncommon for restrictions on planting to be in excess of 100 ft each side of centre line of the overhead electric line. Restrictions on tree planting in proximity to underground cables are also required. This is mostly to avoid the roots of the trees physically disrupting underground cables but also the effect of reducing the moisture content in the ground which has an influence on heat dissipation of the cables.

15.14 Lift and shift provision

It is possible to agree a clause in either a permanent or term easement for the electric line to be relocated. This would either be through a predetermined alternative route or a general provision for relocating the line on other land of the landowner/occupier on a route to be agreed. This would need to provide parameters for agreeing the alternative route to ensure certainty for both parties.

15.15 Development clauses

In certain cases it would be appropriate to agree a clause making provision for the line to be relocated in the event of stated circumstances, for example, the grant of planning consent that would be restricted or prohibited with the presence of the electric line. This would allow an assessment to be made at a future date of the ability to relocate the electric line, either overhead or underground. The alternative would be to provide a payment of compensation based on the value of the land at a later date which fully takes into account the effect of the line inhibiting development. The compensation rules would also apply in assessing land taken and injurious affection[10] caused by the line to the development. Existing agreements often provide for the initial payment provided for in the agreement to be deducted from the compensation payment finally agreed.

---

10   Discussed in sections 15.9 and 15.10.

By way of illustration, if an electric line crossed land in the green belt then an assessment of compensation for the loss of development value would be inappropriate. The method typically employed by licence-holders to assess compensation under a permanent easement, in these circumstances, would be to multiply the annual wayleave payments by a factor of 20. As identified in Chapter 14 these payments essentially compensate for the interference with agriculture operations and the owner's loss at agricultural values. However to accept a compensation payment based on agricultural value in perpetuity denies the owner's true value if planning consent should be granted in the future for a more valuable use. It is therefore appropriate to include a clause providing for the opportunity for the land owner to be paid full compensation in the future for the effect of the electric lines in changed circumstances. Such a clause would not voluntarily be offered by a licence-holder and their preferred approach would be to assess the loss in value applicable at that time. By way of contrast easements for gas pipeline installations invariably pay compensation based on 80% of the freehold value of the land for the easement area affected and also provide a lift and shift provision should planning consent be granted in the future for the land affected by the easement. If a diversion may not be practical then the ability to agree a further payment of compensation in changed circumstances would be an equitable remedy.

15.16 Capitalisation of annual wayleave payments

Where an easement is to be agreed for a term of years then the licence-holder would no doubt look at the annual wayleave payments multiplied by the years purchase for the term agreed at a yield of 5%. While the yield may appear attractive, when compared with property investment historically, this should not be accepted as a matter of fact. This yield has remained constant as it has been applied to agricultural values, upon which the wayleave payments are based. Therefore it would be more appropriate to arrive at the annual loss for the section of electric line concerned by agreeing the lump sum payment, on terms for a permanent easement, and then decapitalise by the appropriate yield for the property. This decapitalised figure would produce an annual value to which would then be applied the years purchase for the term.

15.17 Fibre optic cables

Over recent years, there has been a growth in the extent of fibre optic cables being attached to overhead electricity lines for the transmission of data. The overhead transmission and distribution networks provide routes between major conurbations which can be utilised for telecommunication in one of three ways. These include wrapping a fibre optic cable around the earth wire, replacing the earth wire with an integral fibre optic cable or by attaching a fibre optic cable as a separate catenary between the phase conductors. The rights required for this by a licence-holder are covered in the definition of electric line[11] when used solely for their operational purposes. Where the fibre optic cable is for third party use then a separate permission is required under the Telecommunications Act 1984.

15.18 Blight and sterilisation

Blight has a specific meaning[12] rather than a general description of land affected by unsatisfactory uses. Classifications are specific and therefore limited and include:

- land designated for public authority functions in development land
- land allocated for a public authority function
- land identified for safeguarding land required for the purposes of a public authority
- land declared to be a clearance area under the Housing Act 1985
- land surrounded by or adjoining an area declared to be a clearance area
- land identified to be required as a renewal area
- land allocated for highway purposes
- land allocated for general improvement area under the Housing Act 1985
- land authorised to be required by a private act
- land within a compulsory purchase order including the submission of a compulsory purchase order or a draft compulsory purchase order and includes land before a notice to treat has been served.

---

11 Section 64 of the Act.
12 Defined by schedule 13 to the Town and Country Planning Act 1990.

Unless the licence-holder has prepared and approved a compulsory purchase order under schedule 3 to the Act then blight would not apply to electric lines. Where the electric line is present by virtue of a necessary wayleave the loss cannot be assessed under the compulsory purchase terms as defined for blight. While it may be considered that a property, in the general meaning of the word, is considered to be blighted by an electric line this would not be applied within the assessment of compensation under the Land Compensation Acts. Equally the term sterilised is often applied to land beneath overhead lines where a particular development is thwarted by the presence of the conductors. An example would be a large distribution warehouse where the height of construction plant would contravene the safety requirements of the conductors and the building could not therefore be constructed beneath the line. However, the land could be put to some use, even if not for the original intended use. That could be for landscaping or car parking as provided for within a planning consent. The use may be less advantageous but this would not amount to total sterilisation. If the land beneath the line were considered totally incapable of any form of use, which would be extremely rare, then it would not be unreasonable for a licence-holder to acquire the freehold of that land. However, this would result in potentially a strip of land being removed from a landowner and vested in the licence-holder, who generally only requires the air space above the land to run the conductors. A freehold transfer would therefore be disadvantageous to both parties in those circumstances.

15.19 Surveyors fees

Licence-holders are prepared to pay the reasonable fees of surveyors advising a landowner in entering into an easement, both on the terms and the quantum of compensation. They generally pay a fee based on their own scale of fees[13] published in 1997 and enhanced in 2002. These scales provide a range of schedules to apply to different situations. Where the surveyor has had to assess a claim with a significant element of injurious affection, this requires a greater degree of professional skills and an enhanced scale is therefore chargeable. Where easements are negotiated purely on the basis of a multiplier of

---

13 The Electricity Supply Industry Fee Scales 1997.

the annual wayleave payment, when converting from a wayleave to an easement, then a lower fee basis is set out. To reflect the relatively lower value of agricultural land a separate table provides for higher scale fees to be payable. In all cases a minimum fee is payable. The fee scales take account of situations where a claimant is registered for VAT in order that the appropriate VAT element is claimed from HM Customs and Excise.

15.20 Legal costs

It is also usual practice for a licence-holder to pay the owner's reasonable legal costs in being advised on the terms and implications of entering into a deed of grant and any associated work, for example, in obtaining the consent of the mortgagor and any option holders as well as providing proof of title. However, if the extent of the work is substantially non-contentious conveyancing then the appropriate level of expertise would be reimbursed. This is on the basis of reasonable costs being reasonably required. Any costs incurred by a landowner's legal advisor in objecting to a proposal or discussions not directly relevant to the terms of the deed of grant would not normally be payable.

15.21 Conclusion

It is considered on balance that, certainly for higher voltage lines, the benefit to both landowner and licence-holder in entering into an easement is that it provides certainty. Both parties are clear as to their rights and obligations. For the landowner he will have received full compensation for the loss in value and the licence-holder will have a right enabling it to carry out its business functions and comply with its licence obligations which is a stronger right than a non-statutory wayleave.

Applications to the Lands Tribunal

16.1 Outline of chapter

This chapter explains the procedures which have to be followed if a land owner/occupier is to recover compensation as a result of his land being subjected to a necessary wayleave, a compulsory purchase, at a loss of development value, or an easement. The principles of valuation and the statutory context in which the amount of compensation is calculated are set out in the compensation code contained within section 5 of the Lands Compensation Act 1961 which we have considered in Chapters 12, 13 and 15. A fuller discussion of those principles will be found in *Compulsory Purchase and Compensation* by Barry Denyer-Green (8th ed).[1] This chapter is concerned solely with the procedural aspects of applications of the Lands Tribunal.

16.2 Introduction

The Tribunal is a court of law and a right of appeal on points of law exists from its decisions. Appeals are made directly to the Court of Appeal. This contrasts with the position regarding applications for a necessary wayleave where a decision is made by the Secretary of State based on an inspector's report. No right of appeal exists although possibly there may be a right to seek judicial review of the decision. However, any such considerations are outside the scope of this book.

---

1 See also the footnotes to Chapter 15, section 15.2.

16.3 Lands Tribunal: statutory background

The Lands Tribunal came into existence on 1 January 1950 following the passing of the Lands Tribunal Act 1949 (the 1949 Act). It is the officially designated judicial body whose task is to resolve certain categories of disputes concerning land. Its role and powers cover a wide spectrum of matters many of which are outside the scope of this book. To give the full official brief of the Lands Tribunal, it is to determine questions of disputed compensation arising out of the compulsory acquisition of land; to decide rating appeals; to rule on applications under section 84 of the Law of Property Act 1925 for the modification and discharge of restrictive covenants; and to act as arbitrator on references by consent. Under the 1949 Act other jurisdictions may be added and indeed a number have been. As might be imagined, the Lands Tribunal's duties cover the full spectrum of planning and compulsory purchase issues.

Regulations governing the procedures of the Lands Tribunal inevitably address the full range of the Lands Tribunal's role. To the extent that it is necessary in order to understand the rules for the purposes of this book, we have explained the planning/compulsory purchase background in previous chapters and we reprise the points below, so far as necessary. However, the reader is referred to the many excellent specialist books on planning and compulsory purchase for a proper exposition of those subjects.

16.4 Lands Tribunal: actual working structure

The Lands Tribunal is headed by a president who is the chief judge of the Tribunal. Assisting the President there are members. Cases are usually heard by a single member but they may be heard by two members (where substantial issues of both law and valuation arise) or, in exceptional cases, by three members. The members of the Lands Tribunal will be either lawyers or surveyors. The Tribunal means the President or other members heading the case when it is presented in full.

The Lands Tribunal has a registrar who, together with the deputy registrar and the officials reporting to the tribunal manager, are the persons responsible for all administration of the Lands Tribunal and the majority of the management of the cases.

Hearings prior to the main full hearing and which deal with procedural matters and the preparation of the case are called interlocutory hearings. These are conducted before the registrar of the Lands Tribunal or a single member.

The offices and permanent court rooms are located in London.

The Lands Tribunal states on numerous pages on its website that it will convene hearings and sit outside London if suitable accommodation can be found and whenever the proper disposal of the case requires.

All correspondence to the Lands Tribunal must be addressed to "The registrar".

16.5 Reference to the Lands Tribunal: preliminary groundwork

Where a necessary wayleave is granted to a licence-holder by the Secretary of State under paragraph 6(3) of Schedule 4 to the Act, the occupier of the land and where the occupier is not also the owner of the land, the owner, may recover from the licence-holder compensation in respect of the grant.[2] Where in the exercise of any right conferred by such a wayleave any damage is caused to land or to moveables, any person interested in the land or moveables may recover from the licence-holder compensation in respect of that damage; and if in consequence of the exercise of such a right a person is disturbed in his enjoyment of any land or moveables he may recover from the licence-holder compensation in respect of that disturbance.[3] Compensation under either of the above categories may be recovered as a lump sum or by periodical payments or partly in one way and partly in another.[4]

The compensation reference proceeds on the basis that sections 2–4 of the Land Compensation Act 1961 (the 1961 Act) applies.[5] Section 2 deals with procedural matters. Section 4 deals with costs. We shall consider both sections in detail below when we examine the procedures relating to compensation "references". Before doing that however, other preliminary points need to be made.

Whereas legal action through the courts is begun by the issue of a claim form, in Lands Tribunal matters, a claim for compensation is begun by referring the case to the Tribunal. Similarly, whereas taking legal action through the courts is known as issuing proceedings in

---

2 Paragraph 7(1) of Schedule 4 to the Act.
3 Paragraph 7(2) of Schedule 4 to the Act.
4 Paragraph 7(3) of Schedule 4 to the Act.
5 Paragraph 7(4) of Schedule 4 to the Act.

Lands Tribunal matters of this kind, cases before it are always known as references. The one factor common to both the court and the Lands Tribunal is that the person claiming can still talk about suing for compensation. In the language of the Lands Tribunal the person claiming compensation is the claimant and the licence-holder paying compensation is called the Compensating or Acquiring Authority.

We have already mentioned the regulations governing the Land Tribunal's procedures. These regulations fall into two categories. First there are the Lands Tribunal Rules 1996. These are issued as a statutory instrument under the 1949 Act. We shall refer to these Rules as the 1996 Rules[6] and any reference to a specific rule will follow the convention 1996R[No.]. References to an entire part of the 1996 Rules will follow the convention 1996 Rules Part [No.].

Underpinning the 1996 Rules there are practice directions issued on the authority of the President. The practice directions are designed to explain how the 1996 Rules are to be applied in practical terms in the interests of all concerned.[7] We shall refer to the practice directions by paragraph number following the convention PD[no]. Third, the Lands Tribunal does publish a series of flowcharts giving an overview of the various procedures. We shall cross-refer to those flowcharts where appropriate so that all procedural information both formal and informal is fully integrated in our commentary.

These 1996 Rules and the practice directions will also be relevant to circumstances where the entitlement to compensation arises not from the grant of a necessary wayleave but from the terms of the wayleave itself. They also apply to references arising from the exercise of other statutory powers and to references by consent.

16.6 Questions to be asked before referring a matter to the Lands Tribunal

Before a claimant refers a compensation claim to the Lands Tribunal, two questions need to be asked:

---

6 The 1996 Rules have been amended and we refer to the edition which came into force on 28 April 2006
7 Insofar as any authority is needed for the issue of Practice Directions, the authors suggest that a practice direction is issued by the President pursuant to section 3(6) of the 1949 Act. There is no suggestion in that section that Rules have to be issued by one particular body equivalent to the Rules Committee responsible for the Civil Procedure Rules which apply in the courts.

- which part of the 1996 Rules applies?
- which procedure is most apt for the case?

The answers to these questions must be reflected in the document which the claimant has to lodge with the Lands Tribunal in order to get the reference under way.

16.7 Which part of the rules?

The choice as to which part of the 1996 Rules is between 1996 Rules Part IV (references made under statute) and 1996 Rules Part VII (references by consent). The distinction between the two lies in the procedural rules applicable to the management of the case.

References to the Lands Tribunal in respect of entitlements arising from paragraphs 7(1) and 7(2) of schedule 4 to the Act are clearly references under 1996 Rules Part IV. So too are references arising from compulsory purchase.

The terms of a non-statutory wayleave itself may also provide that if during the existence of the wayleave the landowner suffers a loss in the capital value of his land (usually because development has been prevented) then compensation shall be assessed by the Lands Tribunal. However, actual payment of any compensation so assessed is usually made conditional upon the landowner granting an easement to the licence-holder in substitution for the wayleave.[8] Such a term would give rise to a reference by consent. This would also be the category for references following the grant of an easement voluntarily but under the shadow of compulsion

A reference by consent is conducted by the Lands Tribunal under Part VII as if the provisions of the Arbitration Act 1996 had been incorporated into the 1996 Rules and the practice directions with such adaptations as are necessary.

A consent authorising such a reference may be deemed to have been given long before any actual dispute arises. Let us consider in more detail the kind of term mentioned above which is typical of many to be found in non-statutory wayleaves:

> If at any time during the continuance in force of this Agreement the Grantor considers that the value of the said property has for the purposes of its

---

[8] For the differences between an easement and a wayleave see Chapter 5, section 5.11.

development or use been diminished by reason of the existence of the (electric line) thereon and gives notice thereof to the licence-holder, the licence-holder will pay to the Grantor as compensation an amount equal to such diminution in value such amount in default of agreement between the Grantor and the Licence-holder being determined upon reference by the party to the Lands Tribunal under the Lands Tribunal Act 1949 and the Land Compensation Act 1961. Provided nevertheless that payment of compensation as aforesaid is subject to the Grantor executing in favour of the licence-holder an easement in fee simple free from encumbrances for the electric line across its said property.

The determination of what compensation is payable in these circumstances is certainly a matter over which the Lands Tribunal claims jurisdiction. In its explanatory booklet[9] it states

The Lands Tribunal has power to determine the amount of compensation payable where land is compulsorily purchased by public authority or private body using statutory powers or where land owners claim compensation in respect of depreciation of the value of their land.

Insofar as the dispute arises during the life of an existing non-statutory wayleave and there is within the wayleave itself an agreement to refer, the view adopted by the Lands Tribunal is that this provision would be an arbitration agreement. The fact that parties have made the agreement in the first place is regarded as a sufficient evidence of consent for the purposes of section 1(5) of the 1949 Act and 1996 R25. If the agreement were not construed in that way, then the licence-holder effectively could prevent the reference by withholding consent. This view is supported by section 6 of the Arbitration Act 1996 which defines an arbitration agreement as any "agreement to arbitrate present or future disputes (whether they are contractual or not)"[10] which is designed to catch the widest possible range of clauses. Cross-references to model or standard arbitration clauses or agreements are specifically caught by section 6(2) of the Arbitration Act 1996 provided the cross-reference is a clause of the operative agreement. It is noteworthy in this context and in the context of non-statutory implied wayleaves that to qualify as such an agreement it must be in writing and this in turn is defined as follows:[11]

---

9 Explanatory Leaflet paragraph 3.1.
10 Section 6(1) of the Arbitration Act 1996.
11 Sections 5(2) and (3) of the Arbitration Act 1996.

There is an agreement in writing if:
- the agreement is made in writing (whether or not it is signed by the parties);
- the agreement is made by exchange of communications in writing;
- the agreement is evidenced in writing

...

where parties agree otherwise than in writing by reference to terms which are in writing, they make an agreement in writing.

An agreement is evidenced in writing if an agreement made otherwise than in writing is recorded by one of the parties or by a third party with the authority of the parties to the agreement.

Sections 16.8 to 16.27 inclusive of this chapter are concerned with references under 1996 Part IV. Section 16.28 is concerned with references under 1996 Part VII.

16.8 Which procedure?

The Lands Tribunal will manage a reference under 1996 Part IV in accordance with one of four procedural codes, namely:

- the special procedure
- the simplified procedure
- the standard procedure
- the written representations procedure.

The differences between them lie in the level of case management required before the Lands Tribunal gives it decision. Each procedure will be considered in detail below. Suffice it to say at this stage that which procedure is adopted is ultimately a decision for the Lands Tribunal but the applicant has to give a view when making the reference initially.

16.9 Beginning a claim: preparing a notice of reference

Whether the reference is under 1996 Rules Part IV or 1996 Rules Part VII the reference is begun by the lodging of a notice of reference with the Lands Tribunal.[12] The Lands Tribunal recommends the use of

---

12   1996R9.

Form R (available from the offices or the website). Form R is not obligatory, however, it is designed to enable the applicant to meet the requirements of 1996R10 (which are mandatory). The form comes with extensive notes and the authors recommend the use of Form R as best practice. The notice of reference must be sent with sufficient copies for service upon every other person named in the notice as a potential respondent as well as one for the Lands Tribunal itself.[13] An example of Form R and notes concerning its completion appear in Appendices 22 and 23 respectively. The recommendations concerning compliance with section 4 of the Land Compensation Act 1961 are those of the authors not the Lands Tribunal.[14]

16.10 Beginning a claim: lodging the notice of reference

When Form R has been completed it must be sent with the fee and the right number of copies to the Lands Tribunal.[15] The registrar must then enter particulars of it in the Register of References and then sends a copy of the notice to every party to the proceedings other than the claimant.[16] To state the obvious: the claimant must therefore make and retain his own copy of the signed and dated form.

The registrar then informs all the parties to the reference of the number of the reference and that number becomes the title of the reference for all official correspondence afterwards and in any documents.[17] At this point, the licence-holder (as the Compensating or Acquiring Authority) having received from the Lands Tribunal a copy of the notice of reference and the reference number must acknowledge service and indicate its views on the correct procedure and on experts. There is no specific provision for this in the 1996 Rules or the practice directions. However, operating by analogy with 1996R5(c)(7), the Lands Tribunal gives the Compensating or Acquiring Authority 14

---

13 1996R10(1).
14 The effect of that section which concerns liability for costs in the context of offers made prior to the reference is considered in detail in section 16.14.13 below.
15 1996R11(2) and Flowchart steps 1 and 2.
16 1996R11(1) and Flowchart step 3.
17 1996R11(2) and Flowchart step 4.

days in which to do this.[18] Before continuing our discussion of the process, we need to address two further vital preliminary matters.

16.11 Service

A fuller discussion of the concept of service generally and various statutory provisions relating to it is to be found at Appendix 24. In this section of this chapter we concentrate on the three rules which the Lands Tribunal operates in relation to its own proceedings.

16.11.1 Service of notices

Every party to proceedings shall notify the registrar of an address for service of documents on him.[19] Address for Service is not defined. However, applying by analogy the definition given in the Civil Procedure Rules 1998,[20] an address for service, if not the address of his solicitor, must be the residential address of the party or the address of the party's place of business if that party carries on business within the jurisdiction of the Lands Tribunal (that is to say effectively for our purposes England and Wales[21]).[22] Where a party to proceedings is represented by a person other than a solicitor he shall send to the registrar written authority for that representative to act on his behalf; and notify the registrar if the representative ceases to act on his behalf and, if replaced, shall give the registrar details of the new representative together with the written authority for the representative to act on his behalf.[23]

Any document to be served on any person under these rules shall be deemed to have been served if sent by pre-paid post to that person at

---

18 Flowchart steps 4 and 5. Insofar as any authority is needed for the exercise of this case management power, the authors suggest that PD2 issued under the authority of the President is more than ample for this purpose.
19 1996R54(1).
20 These rules usually abbreviated to CPR are made under the authority of the Civil Procedure Rules Act 1997.
21 The Act does apply the jurisdiction of the Lands Tribunal to Northern Ireland in limited circumstances: section 9 of the 1949 Act and see the discussion in the next sub-section.
22 Commentary at paragraph 6.5.3 of the Civil Procedure Rules 1998.
23 1996R54(2).

his address for service.[24] Note that service is deemed to have been effective if sent by pre-paid post, not if received in the ordinary course of post. It is not a requirement that the pre-paid post be registered. For further discussion of the statutory provisions and other authorities see Appendix 24. Any document to be sent to the registrar under the Lands Tribunal Rules shall be sent to the registrar at the office of the Lands Tribunal.[25] Any application or communication to be made to the President or to any member of the Lands Tribunal in respect of any case shall be addressed to the Registrar at the office of the Lands Tribunal.[26]

16.11.2 Change of address

A party to any proceedings may at any time by notice in writing to the Registrar and to every other party to the proceedings change his address of service under the Lands Tribunal Rules.[27]

16.11.3 Substituted service

If any person to whom any notice or other document is required to be sent under these rules cannot be found after all diligent enquiries have been made; has died and has no personal representative; or is out of the United Kingdom; or if for any other reason service upon him cannot readily be effected in accordance with these rules, the President or the Tribunal may dispense with service upon that person, or make an order for substituted service in such other form (whether by advertisement in a newspaper or otherwise) as the President or Lands Tribunal may think fit.[28]

Note the reference to the United Kingdom in this provision does not determine the jurisdiction of the Lands Tribunal. By section 1(1) of the 1949 Act, the jurisdiction of the Lands Tribunal is for "the remainder of the United Kingdom" excluding Scotland. Application to Northern Ireland is limited by section 9 of the 1949 Act and in any event is irrelevant for the purposes of this book.

---

24 1996R54(3).
25 1996R54(4).
26 1996R54(5).
27 1996R55.
28 1996R56.

16.12 Managing the claim: the overriding objective

The overriding objective is a concept which has been imported from the Civil Procedure Rules 1998 into the 1996 Rules via the Practice Directions. It must be remembered that the Civil Procedure Rules 1998 which govern proceedings in the Court of Appeal, the High Court and the county courts have no application to the Lands Tribunal.[29] Nevertheless when implementing its procedures, the Lands Tribunal will apply similar principles to those underpinning the CPR, for example taking such steps as enable it to deal with cases justly which includes so far as practicable:

- ensuring that the parties are on an equal footing
- saving expense
- dealing with cases in ways which are proportionate to:
  - the amount of money involved
  - the importance of the case
  - the complexity of the issues and
  - the financial position of each party
- ensuring that a case is dealt with expeditiously and fairly; and
- allotting to it an appropriate share of the resources of the Lands Tribunal while taking into account the need to allot resources to other cases.

16.13 Managing the claim to a full hearing: allocation of procedure

Resuming our discussion of the process from section 16.9, the Registrar will then decide which procedure is appropriate to the case.[30] We explain the key features of each procedure in the sections which follow. Where it is necessary to expand on matters which arise during the course of any of the procedures we deal with those in separate sections. The aim of the initial explanation is to give an overview of the path which any reference may take from the lodging of Form R to a full hearing.

---

29 PD2.1.
30 Flowchart step 6.

16.14 Standard procedure

Since the various forms of procedure are variations on the standard procedure we begin with that. The standard procedure comprises steps 7 to 20 inclusive from the Lands Tribunal flow chart.

16.14.1 Standard procedure: basic definition

As the names suggests, the standard procedure is the procedure which applies in all cases for which no specific alternative provision has been made.[31] 1996R29 to 1996R56 inclusive are the rules which together comprise the standard procedure. Under this procedure case management will be in the hands of the registrar. He will look to hold a pre-trial review at the earliest time that it appears appropriate to do so and he will give directions tailored to the requirements of the particular case. These directions may, as appropriate, use elements of the special procedure (for example timetabling through to the hearing date) or the simplified procedure. At any time the registrar or the member to whom the case has been allocated may direct that it should be assigned to one of the other procedures, provided that any consent from a party that is required under 1996R27 or 1996R28[32] has been given.[33]

Any deadline given by the directions or any timetable required by the directions is subject to extension on application[34] to the registrar by any party. However, there is a settled practice and hence there will be a commentary upon the steps that are usually taken.

16.14.2 Statement of case

The claimant will be directed to file and serve a statement of case. The claimant will normally be given 28 days in which to do this.[35] PD8.1 explains that in so ordering the Registrar is adapting the procedure to be found in 1996 Rules Part III which in fact relates to appeals to the Lands Tribunal from the Leasehold Valuation Tribunal. To meet the requirements of 1996 Rules Part III, a Statement of Case must include "full particulars of the facts relied upon and any points of law on

---

31 PD3.6.
32 PD3.7.
33 See the notes on representatives in Appendix 23.
34 1996R35 and 1996R35A.
35 Flowchart step 7.

which the party intends to rely at the hearing".[36] Note the reference to facts. The evidence by which the facts are intended to be proved at the hearing is not to be included in the statement of case.[37] Evidence is usually the subject matter of a separate direction. If not already addressed in the notice of reference,[38] the Claimant should consider to what extent the Statement of Case should be used to demonstrate compliance with Section 4 of the Land Compensation Act 1961.[39]

16.14.3 Reply

Upon service of the statement of case, the acquiring or compensating authority will normally be directed to serve upon the claimant not later than 28 days after the date of the service of the statement of case, a reply.[40] The Reply must consign itself to the "full particulars of the facts relied upon and any points of law on which the party will rely at the hearing". The same points about the exclusion of expert or other witness statements at this stage also applies.

16.14.4 Direction for expert reports and witness statements

On the filing of the Compensating or Acquiring Authority's Reply the parties are directed to file their expert witnesses' reports and witness statements of fact.[41] Expert witnesses testify as to their opinion. Witnesses of fact give evidence of the facts which need to be established if the claim is to be proven or liability is to be reduced or avoided altogether.[42] As the explanatory leaflet makes clear,[43] the aim of such a direction is to prevent parties taking "their opponents by

---

36 1996R8(2) as adapted.
37 Flowchart step 8.
38 See section 16.9 above and Appendix 23.
39 See section 16.14.13 below for a full discussion of the provisions of section 4 of the Land Compensation Act 1961.
40 1996R8(3) as adapted.
41 Flowchart step 9.
42 The law of evidence is itself a substantial topic on which specialist textbooks are written, notably *Phipson on Evidence* (15th ed, Sweet & Maxwell). A more limited discussion relevant to the specific subject-matter of this book is found in sections 16.14.4, 16.14.6, 16.20 and 16.21 below.
43 Paragraph 3.4.

surprise" at the hearing by withholding material until the last minute. It is worth noting that control over "the evidence which may be required or admitted in any reference" is an explicit statutory power given to the Lands Tribunal.[44] The preparation of the evidence is clearly a crucial step. The parties are usually given 28 days in which to accomplish this task. Nevertheless this is the element which is usually the subject-matter of an application for more time.

16.14.5 Interlocutory applications

An application prior to the main hearing is called an interlocutory application. Interlocutory applications are governed by 1996R38. Under the standard procedure, such applications are made direct to the registrar.[45] They must be made in writing and explain the grounds on which the application is made and state the time of the proceedings.[46] No prescribed form is required. A letter to the registrar suffices.

Interlocutory applications can be used for a variety of matters such as: applying to the registrar for an order requiring a preliminary issue to be tried;[47] applying for further information to be given regarding a party's statement of case; an application that specific documents be disclosed. The procedure discussed in this sub-section apply to all forms of interlocutory applications. In relation specifically to an application to extend time, a fee must be paid.[48]

Under 1996R35, the registrar has an unfettered discretion to extend time for the doing of any act or the taking of any step.[49] Such discretion includes granting the extension on such terms as the Registrar thinks fit, even if the deadline has already been missed.[50] PD10 comments that the registrar will require the party applying to justify the application. The explanatory leaflet goes slightly further by stating that "the application must give a full justification for the request".[51] In exercising his discretion the registrar will have regard to the overriding objective. The overriding objective repeats in alternative

---

44 See section 3(6)(8)(iii) of the 1949 Act.
45 1996R38(1).
46 1996R38(2),
47 See section 16.19 below.
48 Explanatory Leaflet Paragraph 9.3.
49 1996R35(1).
50 1996R35(2).
51 Explanatory Leaflet paragraph 9.3.

language the obligation imposed upon the registrar by 1996R38 (7) namely that the registrar "shall have regard to the convenience of all the parties and the desirability of limiting so far as practicable the cost of the proceedings".[52]

Many such applications are made with the consent of the other party, in which case a letter from that other party must be obtained and be sent with the application to the Lands Tribunal.[53]

If the application is opposed, the applicant must send the application (that is to say, the letter to the registrar) to the other party, *before* submitting the letter to the Lands Tribunal. The application shall state that this prior task has been carried out.[54] The letter to the other party ought to be a copy of the actual letter to the registrar, effectively post-dated by a couple of days.

Any party objecting to the application has seven days from the date of the service of the copy letter in which to send written notice of his objection to the registrar.[55]

The registrar *must* consider *all* the objections he has received and may allow any party who wishes to appear before him to do so.[56] The registrar *may* refer the application to the President and *must* do so if either the applicant or the party objecting so requests.[57] The registrar must inform the parties of his decision in writing.[58]

There is a right of appeal to the President against any interlocutory decision. This is exercised by giving notice in writing to the registrar within seven days of the service of the notice of decision.[59] If more than seven days is needed before notice of appeal can be given, the registrar does have power to extend that time.[60] A yet further interlocutory application in the manner described above must be made if such an extension is required. An appeal does not act as a stay of proceedings unless the President so orders.[61]

---

52   1996R38(7).
53   1996R38(3).
54   1996R38(4).
55   1996R38(5).
56   1996R38(6).
57   1996R38(8).
58   1996R38(7).
59   1996R38(9).
60   1996R38(9).
61   1996R38(10).

16.14.6 Documents: 1996R34

The primary issue before the Lands Tribunal (at least, so far as this book is concerned) will be what compensation is due to the claimant, rather than whether compensation is due in the first place.[62] Accordingly, the evidence which the Lands Tribunal will most need to hear is the professional opinion of valuers engaged by each party. Indeed, that is the underlying assumption of the explanatory leaflet[63] which states:

> whilst there is nothing to compel parties to call expert witnesses in support of their cases, they may find that doing so is the only way to establish the merits of the claim. The type of expert witness most commonly called is a surveyor or valuer, but for some cases it is found necessary to call architects, civil engineers, etc. Not more than one expert witness can be called by a party unless they have applied for and been granted permission of the Tribunal[64] except in business disturbance or minerals valuation cases. There is no limit on calling witnesses of fact (that is, witnesses who state what they know)[65] but do not give professional opinions.

There will be annexed to any expert report "whatever plans, valuations, lists of comparable properties, etc that may be appropriate".[66] However, that does not mean that such evidence must be accepted without challenge.

1996R34[67] provides that the Lands Tribunal or subject to any directions given by the Lands Tribunal the registrar may, on the application of any party to the proceedings or of its own motion, order any party:

- to deliver to the registrar any document or information which the Lands Tribunal may require and which it is in the power of the parties to deliver
- to afford to every other party to the proceedings an opportunity to inspect those documents (or copies of them) to take copies

---

62   See sections 16.5 and 16.7 above. Remember that a necessary wayleave hearing does not assess compensation but the grant of a necessary wayleave creates a statutory right to be compensated: see paragraph 7 of schedule 4 to the Act.
63   Explanatory leaflet paragraph 9.1.
63   See section 16.21.
65   In relation to the scope of factual evidence see section 16.19.
66   Explanatory leaflet paragraph 9.2.
67   1996R34(1).

- to deliver to the registrar an affidavit or make a list stating whether any document or class of document specified or described in the order or application is or has at any time been in his possession, custody or power and stating when he parted with it
- to deliver to the registrar a statement in the form of a pleading setting out further and better particulars for the ground on which he intends to rely and any facts or contentions
- to answer interrogatories on affidavit relating to any matter at issue between the claimant and the other party
- to deliver to the registrar a statement of agreed facts, facts in dispute in the issue, or issues to be tried by the Lands Tribunal or
- to deliver to the registrar witness statement or proofs of evidence.

The language of 1996R34(1) mirrors the language of the Supreme Court Practice prior to the reforms introduced by the Civil Procedure Rules 1998. An affidavit is a written statement made on oath.[68] The person who makes an affidavit, that is to say the one who deposes to the facts and matters set out in the affidavit is called the deponent.[69] The statement is prepared by the party's own solicitor. However, it is formally signed by the deponent in front on an independent solicitor. The deponent has to take an oath on an appropriate piece of scripture according to the deponent's religious convictions and make a formal statement before the independent solicitor can witness the signature on the document. Further and better particulars means further details as to the nature of the grounds on which a party intends to rely. Interrogatories are formal questions put to a party in connection with a case. The word interrogate in this context having significantly less sinister meaning than the modern understanding of the word. Further and better particulars and interrogatories are now dealt with under the Civil Procedure Rules, Part 18 under the general heading of Requests for Further Information.

Where an order is made under 1996R34(1) the Lands Tribunal registrar may give directions as to the time within which any document is to be sent to the registrar (being at least 14 days from the date of the direction) and the parties to whom copies of the document are to be sent.[70] The time-limit is capable of extension as previously discussed.

---

[68] It comes from the Medieval Latin for "he has sworn".
[69] Depose and deponent in their strict legal sense are derived from the Latin verb meaning to commit something to writing in solemn form.
[70] 1996R34(2).

The rule as to interlocutory applications[71] shall apply to any application under this rule as appropriate, both in relation to applications made by one of the parties or to any matter where the registrar acts of his own motion, that is to say on his own initiative without being prompted by an application by either party.

For the reasons explained when we discuss these procedures below, it would rarely be appropriate to resort to 1996R34 if the reference has been allocated, either to the simplified or to the written representations procedure. However, it is a valuable tool in references allocated to the standard or to the special procedure.

16.14.7 Listing questionnaires

Assuming all the evidence has been served and filed, and the other directions have been observed, the Lands Tribunal sends listing questionnaires to both parties asking for their available dates for the hearing, preferred venue for information any special arrangements for access or video conferencing which may be required and finally their time estimate for the duration of the hearing.[72] The parties will be given 14 days to reply to the questionnaires.[73] If a party does not reply, the Lands Tribunal may list the hearing at a venue and on dates that may not be convenient to the party.

The parties are also required to inform the Lands Tribunal whether any other application is contemplated which might affect the listing of the hearing, and whether issues have been narrowed by agreement and if so what the issues are which the Lands Tribunal must determine. Finally, the Lands Tribunal asks about the practical arrangements for the lodging of the agreed bundle of documents for the hearing. If any of these matters are unresolved a pre-trial review may be fixed.

The purpose of the listing questionnaires is to enable the Lands Tribunal to have all relevant information to fix a hearing, not only at an appropriate venue, but also within the appropriate time estimate so that the necessary judicial resources can be allocated. This is why the parties are required to tell the Lands Tribunal how long the hearing is expected to take.[74] It is good practice for the advocates to supply this information. They should consult with each other about this and agree a time

---

71 1996R38.
72 Flowchart Step 11.
73 Flowchart Step 12.
74 Explanatory leaflet paragraph 10.2.

estimate. If the time estimate is wrong, then they may have to have an inconvenient, possibly expensive, adjournment part way through the case until more available days can be found. The explanatory leaflet makes it plain that it is important in the interests of all litigants that the resources of parties and of the Lands Tribunal should not be wasted by unnecessary adjournments or by over-estimates of the days required.

The Lands Tribunal then fixes a hearing. The practice of the Lands Tribunal where counsel has been briefed to appear and counsel's availability has been notified via the listing questionnaire is to contact counsel's clerk directly about hearing arrangements.[75] If counsel has not been briefed to appear then the Lands Tribunal communicates with the parties directly or through their other representatives. The Lands Tribunal identifies a date for the hearing. The parties are notified of the date(s) and venue. The parties must inform their witnesses.[76]

16.14.8 Documents: PD14

In cases assigned to the standard procedure, unless the Lands Tribunal directs otherwise, the parties should lodge with the Lands Tribunal not less than 14 days[77] prior to the hearing sufficient copies (for the number of members sitting) of a fully paginated agreed trial bundle containing the following:

- expert witness reports, including all appendices, photographs and plans referred to
- witness statements
- all other documents to be relied upon
- a statement of agreed facts and issues.

The advocates' skeleton arguments (that is to say an outline of their arguments both on matters of fact and law) should be lodged not less than seven days before the hearing. Photocopying of any cases relied upon by the advocates should be provided for the Lands Tribunal.[78]

Plans and photographs should be appropriately annotated and indexed. Plans should be in A4 or A3 format, unless there is a good reason to use another size.

---

75 Explanatory leaflet paragraph 10.3.
76 Flowchart Step 13.
77 Flowchart Step 14.
78 PD14.1.

It should be noted that this is a statement of that which is agreed and a bundle of the documents which are agreed. If it is not possible to agree certain facts and issues then it is necessary to explain to the Lands Tribunal the extent to which such matters are not agreed. The purpose of this practice direction is not to exclude from the hearing any evidence on matters which remain in dispute between the parties. The practice direction nevertheless recognises that where the bulk of the evidence is that of an expert rather than of a witness to fact, there will be matters on which the experts are able to agree. The title to the land in question and its area are two obvious examples.

16.14.9 Final hearing

Subject to the 1996 Rules and to any direction by the President, the procedure at the hearing of any proceedings shall be such as the Lands Tribunal may direct.[79] The registrar and the Lands Tribunal have power to administer oaths and take affirmations for the purpose of affidavits to be used in proceedings or for the purpose of giving oral evidence at hearings.[80] Evidence at the final hearing will invariably be given on oath or affirmation. An affirmation is a solemn declaration that the evidence will be true, but does not have the religious connotation of the oath.

The custom and practice is for the claimant to begin the hearing by a statement setting out the case and then calling evidence and presenting the documents, which at a hearing are called exhibits.

The evidence which any witness (whether expert or of fact) gives first is called evidence in chief. Originally all such evidence would be given by the witness in response to open questions presented to the witness by counsel for the party by whom the witness had been called to court. The purpose of the provision of expert reports and witness statements prior to the hearing is to shorten this process very considerably. An expert's report will stand as the expert's witness statement unless the Lands Tribunal directs otherwise. A witness statement will stand as the witness' evidence in chief.[81] However, a witness giving oral evidence may, with the consent of the Lands Tribunal, (a) amplify his witness statement and (b) give evidence on new matters which have arisen since the witness statement was

---

79 1996R48.
80 1996R41.
81 PD15.2.

served. Notice of any such additional evidence should, where possible, be given to the other party and any failure to do so will be taken into account by the Lands Tribunal in deciding whether to give consent to such evidence being tendered at the hearing.

Questions put to the witness by the opposing party's advocate are called cross-examination. From time to time the witness will then be questioned by the original counsel on any matters which have arisen out of the cross-examination. This is called re-examination.

After the claimant has called his witnesses and presented his evidence the licence-holder then has the opportunity to set out its own case and call its own witnesses. The hearing will conclude with a speech on behalf of each of the parties, setting out the legal arguments on which it relies in support of its case and proposing to the Lands Tribunal the conclusions which the Lands Tribunal ought to draw from the evidence. The final hearing then closes.

16.14.10 Issue of decision

The Lands Tribunal rarely gives its decision immediately, but rather reserves it for consideration.[82] Subject to one exception, the decision of the Lands Tribunal on a reference shall be given in writing and shall state the reasons for the decision.[83] The exception is that the Lands Tribunal may give its decision orally at the end of the case where the Lands Tribunal is satisfied that this would not result in any injustice or inconvenience to the parties.[84] However, the Lands Tribunal makes clear that an oral decision at the end of the actual hearing is not normally appropriate.[85]

The registrar is obliged to serve a copy of the decision on every party who has appeared before the Lands Tribunal in the proceedings.[86] Where an amount awarded or value determined by the Tribunal is dependent upon the decision of the Lands Tribunal on the question of law which is in dispute in the proceedings, the Lands Tribunal shall ascertain and shall state in its decision any alternative amount or value which it would have awarded or determined if it had come to a

---

82   Flowchart Step 15.
83   1996R50(1).
84   1996R50(2).
85   PD20.
86   1996R50(5).

different decision on the point of law.[87] The purpose of this is to enable the Court of Appeal to enter an alternative judgment if the appeal is successful, without forcing the parties to go back to the Lands Tribunal and re-argue the case.

16.14.11 Formal conclusion of the reference

At this point the parties, while having a full reasoned decision from the Lands Tribunal do not actually have a formal order stating the effect of the decision. Such an order is necessary if, for example, further action has to be taken in the courts following the Lands Tribunal decision or in order to recover costs. The 1996 Rules[88] provide that the Lands Tribunal may, and on the application of any party to the proceedings, shall, issue an order incorporating its decision. In reality such an order is invariably made without a specific request from one or other of the parties.

If no costs award is made then the reference concludes at this point, although the hearing fee must be paid.[89] Under the Lands Tribunal (Fees) Rules 1996 the hearing fee is 2% of the amount awarded or determined, subject to a minimum fee of £100 and a maximum fee of £5,000. The Lands Tribunal does have power to reduce or remit fees in the case of hardship.[90] Unless the Lands Tribunal directs otherwise, the appropriate hearing fee is payable by the claimant, although the claimant's right to recover the hearing fee from the acquiring or compensating authority as part of an order for costs is not thereby affected.[91] A solicitor acting for a party must be on the record and he will be responsible for the fees payable by that party while he is on the record.[92]

16.14.12 Award of costs: procedure

On the issue of the main decision the parties are then invited to send in written submissions as to who should bear the costs of the case.[93]

---

87 1996R50(4).
88 1996R50(3).
89 Flowchart Step 19.
90 PD21.
91 PD21.
92 1996R53.
93 Flowchart Step 16.

Cost submissions must be filed within 14 days of the request.[94] The principles regarding costs are discussed in section 16.14.13 below.

The Tribunal considers the submissions received and makes a decision on the costs of the case based upon the written submissions. The costs decision is incorporated as an addendum to the main decision and sent to the parties. The decision takes effect from this point.[95]

The registrar may make an order as to costs in respect of any application or proceedings heard by him.[96]

16.14.13 Award of costs: principles

There is no right to costs. Indeed there are certain restrictions on the right to recover costs from an opponent.

First, the effect of section 4 of the 1961 Act on references in relation to compensation caught by that at section is that a statutory restriction on the quantum of costs recoverable from any reference is imposed. The policy of the provision is to encourage the making and acceptance of offers prior to the lodging of a reference. If a licence-holder has made an unconditional offer of compensation which exceeds the amount which the claimant is awarded at the conclusion of the reference[97] or if the Lands Tribunal is satisfied that the claimant has failed to deliver a notice in writing of the amount claimed with the particulars prescribed,[98] to the licence-holder in sufficient time to enable the licence-holder to make a proper offer,[99] then in the absence of special reasons which lead the Lands Tribunal to think it improper so to do, the Lands Tribunal shall order the claimant to pay those costs incurred after date on which the offer was made or after the date on which a claimant's notice of claim should have been made. If the claimant does deliver a notice in writing containing the prescribed particulars and he makes an unconditional offer which the licence-holder fails to beat at the reference,[100] then again in the absence of special reasons which would make it improper to do

---

94   Flowchart Step 17.
95   Flowchart Step 18.
96   1996R52(2) subject to the right of appeal to the President contained in 1996R52(3).
97   See section 4(1)(a) of the Land Compensation Act 1961.
98   The relevant particulars are set out in section 4(2) of the Land Compensation Act 1961.
99   See section 4(1)(b) of the Land Compensation Act 1961.
100 See section 4(3) of the Land Compensation Act 1961. "Fails to beat" is a paraphrase of the point that the claimant need only equal the amount of the pre-reference offer.

so, the Lands Tribunal shall order the licence-holder to bear its own costs and pay those of the claimant from the date of the offer. In any event the Lands Tribunal can always disallow the costs of counsel.[101] Where the claimant is ordered to pay costs in the exercise of these powers the costs can either be deducted from the compensation awarded[102] or recovered summarily as a civil debt.[103]

Second, subject to that statutory restriction, the costs of and incidental to any proceedings shall be in the discretion of the Lands Tribunal.[104]

Third, there are specific rules applicable to the simplified procedure which we consider below.

Subject to each of those preliminary considerations, the Lands Tribunal will normally award costs on the standard basis.[105] On this basis, costs will only be allowed to the extent they are reasonable and proportionate to the matters in issue and any doubt as to whether the costs were reasonably incurred or reasonable and proportionate in amount will be resolved in favour of the paying party. Exceptionally the Lands Tribunal may award costs on an indemnity basis. On this basis the receiving party will receive all his costs, except for those which have been unreasonably incurred, or which are unreasonable in amount. Any doubt as to whether the costs were reasonably incurred or are reasonable in amount will be resolved in favour of the receiving party.

The ability to make either a summary or a detailed assessment is expressly permitted by the 1996 Rules.[106] Further, the ability to direct that costs be assessed on either the standard or the indemnity basis is likewise expressly permitted.[107] It should be noted that when comparing the practice directions with the 1996 Rules themselves, the 1996 Rules have been written in language which preceded the Civil Procedure Rules 1998. Accordingly, assessment is referred to by the former technical term of taxation. If costs were reduced upon assessment by the court they were regarded as being taxed down. A sum not allowed and struck out of the bill was a sum taxed off. The practice directions are written in language more consistent with the Civil Procedure Rules 1998.

---

101 See section 4(4) of the Land Compensation Act 1961.
102 See section 4(5) the Land Compensation Act 1961.
103 See section 4(6) of the Land Compensation Act 1961.
104 1996R52(1).
105 PD22.8.
106 1996R52(4).
107 1996R52(4).

16.14.14 Detailed assessment of costs

If a costs order is made following the presentation of the written submissions on costs the parties will be sent a copy of the Tribunal's detailed assessment of costs procedure flowchart which explains the detailed assessment procedure. This will be closely mirrored on the provision for the assessment of costs under the Civil Procedure Rules. Detailed assessment refers to an in-depth examination of the costs of the whole of the reference at the end of the case.[108] It is to be contrasted with summary assessment which is an assessment which the Lands Tribunal can make, either in a simple case or at an interlocutory hearing. A party who proposes to apply for a summary assessment should prepare a summary of the costs and should serve it in advance on the other party. Costs of a trial which are to be the subject of a detailed assessment are referred to the registrar.[109]

The process of a detailed assessment is that the party with the benefit of the order or costs has to draw up a detailed bill, setting out the cost of each item of work for which a repayment is claimed and the nature of the work done. The registrar considers the bill item by item and decides whether to allow or disallow each item. The fee for an assessment is £0.05p for every £1 of the total bill permitted on the assessment.[110]

If the receiving party is dissatisfied with the decision of the registrar on an assessment, he may, within seven days of receiving notice of the outcome of the assessment, serve on any other interested party and on the registrar written objections specifying the items objected to and applying for the assessment to be reviewed in respect of those items.[111] Upon such application the registrar shall review the assessment of the items to which objection has been taken and shall state in writing the reasons for his decision.[112]

A person dissatisfied with the decision of the registrar upon a review may, within 10 days of the decision, apply to the President to review the assessment and the President may make such order as he thinks fit, including an order as to payment of the costs of the review.[113]

---

108   Flowchart Step 20.
109   1996R52.
110   Paragraph 13.1.
111   1996R52(5) and PD22.8.
112   1996R52(6) and PD22.8.
113   1996R52(7) and PD22.8.

An application for a review of an assessment of costs is made in exactly the same form as an interlocutory application.[114]

16.15 Special procedure

The special procedure adapts the standard procedure in two significant respects. On allocation the case management is placed in the hands of a member of the Lands Tribunal, rather than the registrar. This will occur if the case is sufficiently complex or there is a significant amount in issue or it has a wider importance.[115]

Once a case has been allocated to a member or members under the special procedure, the members will order a pre-trial review and the parties are notified accordingly.[116] The allocation of the case to a member for case management does not alter the fact that all communications continue to be addressed to the registrar, who indeed will continue to have some involvement in the administrative support for, and the management of, the case.

Pre-trial reviews are governed by 1996R39 and are considered in a separate section below. The use of one or more pre-trial reviews is the particular hallmark of the special procedure. The member gives a full programme of directions tailored to the case in question. They will include directions of the parties to file and serve a statement of case and a reply.[117]

Subject to any particular directions the member gives at the pre-trial review, the procedure followed thereafter is broadly identical to the standard procedure discussed above.[118]

16.16 Simplified procedure

The simplified procedure is the name given to the procedure authorised by 1996R28. The purpose of this procedure is to provide for the speedy and economical determination of cases in which no substantial issue of law or valuation practice or substantial conflict of fact is likely to arise. It is often suitable where the amount at stake is small.[119]

---

114 1996R52(8).
115 PD3.2.
116 Flowchart Step 7a.
117 Flowchart Step 7b.
118 Flowchart Step 7c.
119 PD3.3.

The simplified procedure can only be invoked if the claimant so consents.[120] The simplified procedure begins when a member or the registrar directs that proceedings shall be determined in accordance with 1996R28. Remember that the claimant will have had to express observations on the appropriate procedure in making the reference in the first place using Form R.

The registrar shall send a copy of the simplified procedure direction to all the parties of the proceedings and any party objects to the direction may, within seven days of the service of the copy on him, send written notice of his objection to the registrar. The registrar is then obliged to reconsider his decision.[121] In dealing with any objection the provisions of interlocutory applications apply as appropriate.[122]

If a request to reconsider is made the registrar considers it and the other party's comments. If he rescinds the simplified procedure direction, the standard procedure will be followed. If no request is made or if the Registrar confirms a simplified procedure after a request is made, then the simplified procedure takes effect.[123] Whereas under standard and special procedure, the hearing date is only fixed after the submission of listing questionnaires and therefore after compliance with directions previously made, under the simplified procedure the fixing of the hearing date and the notifying of the hearing date to the parties is actually the first step that occurs.[124] The remaining directions are all geared to preparing the case for hearing on that hearing date.

The claimant will be directed to file and serve a statement of case. The acquiring or compensating authority will be directed to file and serve a reply. In each case there will be an interval of 28 days for each party to accomplish that task.[125] 1996R28(6) which authorises these directions requiring a minimum of 21 day's notice of the day fixed for the hearing of the proceedings to be given.

Not less than 14 days before the hearing date the parties must send to each other copies of all the documents on which they intend to rely.[126] Not less than seven days before the hearing date each party must send to the other party copies of any expert witness' report on

---

120 1996R28(1).
121 Flowchart Step A.
122 1996R28(3) applying 1996R38 (6) to (9) and (11) as appropriate.
123 Flowchart Step B; 1996R(4).
124 Flowchart Step C.
125 Flowchart Step D .
126 Flowchart Step E; 1996R28(7)(a).

which they wish to rely and a list of the witnesses whom they are intending to call at the hearing.[127]

The registrar does have power from time to time, whether on the application of a party or his own motion, to amend or add to any direction issued if he thinks it necessary to do so in the circumstances of the case.[128] However, the whole aim of the simplified procedure is to avoid additional applications and to produce as streamlined a process leading to the final hearing as possible.

Not less than seven days before the hearing date the party who lodged the reference must file a statement of agreed facts and issues with the Lands Tribunal.[129] The final hearing takes place — almost always being completed in a single day.[130] It is a special feature of a simplified procedure hearing that the hearing shall be informal and shall take place before a single member of the Lands Tribunal who shall act as if he were an arbitrator and who shall adopt any procedure that he considers to be fair.[131]

Strict rules of evidence shall not apply to the hearing and evidence shall not be taken on oath unless the Tribunal orders otherwise.[132] The member hearing the case usually gives his decision later in writing.

Except in compensation cases governed by the cost provisions in section 4 of the 1961 Act, an order for costs will not normally be made in simplified procedure cases.[133] The exception to this rule is that the Lands Tribunal may make such an award in cases where an offer of settlement has been made by a party and the Lands Tribunal considers it appropriate to have regard to the fact that such an offer has been made or, the Tribunal otherwise regards the circumstances as exceptional. If exceptionally an award of costs is made, the amount shall not exceed that which would be allowed if the proceedings had been heard in a county court.[134]

---

127 Flowchart Step F; 1996R28(7)(b).
128 1996R28(8).
129 Flowchart Step G — this is not specifically required by 1996R(28) but there is clearly power to make such a direction under 1996R34(1)(f) as an exercise of the provision given by 1996R28(8).
130 Flowchart Step H.
131 1996R28(9).
132 1996R28(10).
133 1996R28(11).
134 1996R28(11)(a) and (b).

If a costs order is made, then again the costs are subject to a detailed assessment and the detailed assessment of costs flowchart will be supplied to the successful party.

At any time if a party so applies using the same provision as an interlocutory application or if the Lands Tribunal on its own initiative so decides, a case can be taken out of the simple procedure if that appears appropriate and in that event the Lands Tribunal may give directions for the disposal of the proceedings by some other form of procedure.[135]

16.17 Written representations procedure

Sections 6A to 6C inclusive[136] of the 1949 Act permit the Lands Tribunal to determine cases without an oral hearing, subject in the case of a reference for compensation to the consent of the claimant. The written representations procedure is a permitted derogation from the rule that hearings be in public which section 2(2) of the Land Compensation Act 1961 otherwise imposes upon references to the Lands Tribunal.[137] 1996R27 deals specifically with the Rules relating to such a procedure, which is called the written representations procedure. 1996R27(1) goes further than the express provision in the 1949 Act by requiring that all the parties to the proceeding must agree to the reference being determined without an oral hearing. Such an application may be made at any time during one of the other procedures to which the reference may have been allocated.[138] The Tribunal determines the application as it would determine any other interlocutory application and then makes direction confirming whether or not the representations procedure is to be used.[139]

The Tribunal is then obliged to give directions relating to the lodging of documents and representations as it considers appropriate.[140] If the Tribunal refuses the application the reference continues under one of

---

135 1996R28(12).
136 Inserted into the 1949 Act by the Local Government, Planning and Land Act 1980, section 193 and paragraphs 3(1) and 3(3) of Schedule 33 to that Act.
137 The proviso to section 2(2) was inserted by the Local Government, Planning and land Act 1980, section 193 and paragraph 5 of Schedule 33 to that Act.
138 Flowchart Step WR1.
139 Flowchart Step WR2.
140 1996R27(3).

the other procedures.[141] Assuming however a direction for the application of the written representations procedure, the claimant will normally be given 28 days to file and serve his or her written representations.[142] Oddly 1996R27(2) suggests that a party to the proceedings has a discretion whether or not to submit representations to the Lands Tribunal. However if oral hearing has been avoided it is difficult to imagine a party not wishing to exercise that right! The licence-holder files and serves its written representations within the time directed, normally within 28 days of service of the claimant's written representations.[143] The claimant files and serves a reply to the licence-holder's written representation. The claimant will normally be given 14 days to do so.[144] Once the parties have served and submitted their written representations to the Lands Tribunal, it will then determine the reference by issuing its decision. The parties may then have the opportunity to make a submission as to costs as if step 17 onwards from the standard procedure applied.

The Lands Tribunal indicates[145] that the written representation procedure will be followed only if the Lands Tribunal having regard to the issues in the case and the desirability of minimising costs is of the view that oral evidence and argument can properly be dispensed with. The determination of liability on the basis of written representations alone is a common feature, for instance, in the field of rent review. If the reference depends upon relatively straight forward valuation evidence which does not need to be tested by cross-examination, then certainly the written representations procedure would seem to be the most obvious way of obtaining a decision with the minimum of cost. It will often only be possible to approach the opposite party for a consent to agree to the written representations procedure if there has been an informal exchange of evidence prior to the making of the reference. Such pre-action exchanges would certainly be encouraged in the furtherance of the overriding objective.

---

141 Flowchart Step WR3.
142 Flowchart Step WR4.
143 Flowchart Step WR5.
144 Flowchart Step WR6.
145 PD3.5.

16.18 Pre-trial review

A pre-trial review is a hearing prior to trial which occurs in a reference. The parties are required to attend.[146] They are given not less than 14 day's notice of the place, date and time of the hearing.[147] A pre-trial review is the particular distinguishing feature of the special procedure. However, it is worth remembering that either party may request one. Indeed the first question on the listing questionnaire asks the parties to say whether one is needed.

It is not the purpose of the pre-trial review to make standard directions in the absence of the parties but rather to consider what directions are necessary or desirable for securing the just, expeditious and economical disposal of the reference.[148] It is essential, therefore, if the parties are to be properly prepared that there should have been an attempt to explore the issues in correspondence beforehand to see what measure of agreement might be reached. Indeed the Lands Tribunal expects the parties so far as they are able to do so to inform it as to these matters prior to the date of the review.[149] Not only must the issues be identified so far as possible but also the number of proposed expert witnesses, their areas of expertise and the scope of their evidence is also information which the Lands Tribunal requires. If a party is not able to give this information then an explanation of why not will certainly be needed.

It follows from this that a thorough review by each party of the details of the case is a clear prerequisite. It is best practice to conduct this review with the advocate who will conduct the reference at the final hearing. Moreover, it is equally good practice to be represented at the pre-trial review by that advocate. This is because it is also part of the mandate of the Lands Tribunal at the pre-trial review to secure as far as they can that the parties make all such admissions and agreements as ought reasonably to be made by them in the proceedings[150] and to record an admission or agreement and any refusal so to admit or agree in the order.[151] For obvious reasons a party will resist making an admission or agreement which might harm the case, but any unreasonable refusal will expose that party to an adverse

---

146   1996R39(1).
147   1996R39(2).
148   1996R39(3)(a).
149   PD3.2.
150   1996R39(3)(a).
151   1996R39(3)(b).

costs order if the matter which could have been admitted or agreed is subsequently proved at the final hearing. Whether a party has acted reasonably in the conduct of proceedings is a matter which the Lands Tribunal specifically takes into account on the question of costs.[152] It is the purpose of a pre-trial review to shape the reference. It is therefore desirable for the advocate who must ultimately present the case to participate in that process.

Where appropriate, a date for the hearing will be fixed at the pre-trial review and the directions and the dates for subsequent pre-trial reviews will be scheduled by reference to this date.[153] It is therefore crucial that dates for future availability of the key witnesses and the advocate are brought to the pre-trial review.

Where a party seeks a specific direction he ought so far as he can to make an application for that direction at the pre-trial review and give notice in writing both to the registrar and to the other parties of its intention to do so.[154] Failure to take advantage of the pre-trial review in this way can result in costs penalties being incurred. If a party makes at a later date an application which might have been made at a pre-trial review then unless the Lands Tribunal considers that there was sufficient reason for the failure to make the application at the pre-trial review then the person applying must pay the costs of and occasioned by the application even if the order sought is made.[155] Note, this sanction does not depend upon a person behaving unreasonably. A mere oversight is enough unless there is sufficient reason to excuse it. This is an important incentive to engaging in a sensible exchange of views with the other parties. You cannot be clear about what directions might be necessary unless you have a full picture not only of your own case but also the opposite party's. It should also be borne in mind that a solicitor for a party could be put into an invidious expedition if the "sufficient reason" is that his client is either failing to give instructions or failing to pay interim accounts. The disclosure of such information as the price of forwarding an adverse costs order may be a high price to pay if it destabilises the case from a tactical point of view.

Applications at a pre-trial review are made in the same way as any other interlocutory application.[156] The giving of notice of any

---

152 PD22.2.
153 PD3.2.
154 1996R39(4).
155 1996R39(5).
156 1996R38(6).

application and the consideration of any objections to such application are therefore an integral part of the pre-trial review. However in connection with the pre-trial review, there is an added sanction. If any party does not appear at the pre-trial review the Lands Tribunal or the registrar may, after giving the parties the opportunity to be heard make such order as may be appropriate for the purpose of expediting or disposing of the proceedings.[157] "Disposal" means a final order. In other words a failure to attend exposes the absent party to an award in default being made against him.

16.19 Preliminary issues

A preliminary issue is a particular issue in a case which is heard by the tribunal separately from all the other issues and on which the Tribunal delivers a binding judgment. It has the potential to be decisive of the whole case, or at least spare the parties considerable time and cost in preparing for a hearing on such other issues as may remain to be determined after the preliminary issue has been dealt with.

The President or the Lands Tribunal may on the application of any party to proceedings order any preliminary issue in the proceedings to be disposed of at a preliminary hearing.[158] Any such application has to be made using the procedure for an interlocutory application discussed above.[159] Although it is a matter for the parties to consider whether or not to apply for an order directing that a preliminary issue be heard, the Lands Tribunal makes it clear that it will pro-actively seek to draw attention to issues which in its view might usefully be determined as a preliminary issue with a view to saving time and expense.[160]

The Lands Tribunal does not insist that a preliminary issue be decisive of the whole case. It will be sufficient if there will be a notable reduction in the issues of the case with the associated costs and delay associated with the preparation of a hearing on those issues and in the length of the hearing itself. Obviously it is important not to propose a hearing into what is said to be a preliminary issue if in fact the Lands Tribunal can only deal with the issue, having heard all the evidence and the arguments that would have been tendered at a main hearing

---

157   1996R38(7).
158   1996R43 (1).
159   1996R43 (3).
160   PD9.1.

anyway. The crucial distinction is whether the preliminary issue can be effectively separated or severed from all the other issues in the case.[161]

An application under 1996R43(1) should set out with precision the point of law or other issue or issues to be decided. It should, where appropriate, be accompanied by a statement of agreed facts and it should state whether in the view of the party making the application the issue can be decided on the basis of the statement of agreed facts or whether evidence will be required. If evidence is said to be needed the application should state what matters that evidence would cover. The application should state why, in the claimant's view the determination of the issue as a preliminary issue would be likely to enable the proceedings to be disposed of more expeditiously and/or at less expense.[162] Such an application is clearly different in nature and extent from many other interlocutory applications. It does not assist the Lands Tribunal if on the actual hearing the advocate presenting the application seeks to amend the wording of the preliminary issue because on further reflection it does not actually meet the requirements of the case. Since the letter by which the application is made must set out grounds of the application at some length, best practice would indicate that the terms of the letter, not just the wording of the preliminary issue, be discussed with the proposed advocate before it is sent to the Registrar and to the opposite party. Invariably preliminary issues will only be identified as having the potential to be heard separately after an exchange of correspondence between the parties where the issues are candidly and honestly debated.

The Lands Tribunal reserves the right to require that witness statements and documentary evidence be filed with it before it can consider an application for a determination of the preliminary issue. If it decides to exercise that prerogative then directions will be made for the preparation of the hearing of the application. This will include directions as to the filing in advance of any experts' reports, witness statements, documentary evidence and such further statements of agreed facts that appear to it to be required.[163]

Whether or not an order for any preliminary issue is made, the hearing of the application can itself be a useful pre-trial investigation of the merits and strengths and weaknesses of the claim. Two best practice

---

161 PD9.2.
162 PD9.3.
163 PD9.4.

recommendations arise from this. The first is that the advocate who will present the final hearing should be present at the hearing of the application for the preliminary issue. Second, there should be a full attendance from the clients as well as the legal or other professional team. It is quite possible after such a hearing that settlement discussions may be started or resumed.

16.20 Witness of fact evidence

Those giving evidence at a hearing, as a witness as to fact, should provide a written statement verified by a statement of truth. The form of a statement of truth is as follows: "I believe that the facts stated in this witness statement are true".[164]

There are no practice directions further expanding upon the form or content of a witness statement. Good practice can found in the principles behind the production of witness statements which are contained in the Civil Procedure Rules 1998. The subject of evidence is dealt with in Part 32 and the practice direction supplementing it, the key points of which are set out below.

- A witness statement should be confined to evidence that the person making it would be allowed to give orally and, as hearsay evidence is generally admissible in civil proceedings, may include such evidence.
- The witness statement must, if practicable, be in the intended witness' own words and the statement should be expressed in the first person and should also state details of name, place of residence, occupation and any relationship to the party in the proceedings. The document should clearly identify that it relates to a particular reference by including the title of the reference at the heading and also indicating which witness statement it is and whether any exhibits are intended to annexed to it.
- The witness statement indicates which of the statements are made from the witness' own knowledge and which are matters of information or belief and the source of any matters of information or belief.
- Witness statements should contain evidence which proves the facts which have to be established at the hearing in order to make the

---

164 PD15.1.

case. They should not be used as a vehicle for argument (especially not legal argument). Nor should they be used to comment on each individual document which has been disclosed in the proceedings.

16.21 Expert evidence

16.21.1 Introductory overview

1996R42 permits a party in any proceedings before the Lands Tribunal to call expert evidence. The expectation of the Lands Tribunal is that it will be called upon to hear and evaluate the evidence of experts in most of the references before it.[165] Expert witnesses are defined as those who are qualified by training and experience in a particular subject or subjects to express an opinion on any issue which the Lands Tribunal is considering.[166] This is the crucial difference between an expert witness and a witness as to fact. Witnesses as to fact are not supposed to express opinions. Most frequently the expert opinion evidence will relate to a valuation or surveying matter. However, the principle applies equally to any other professional discipline. For instance, if a claim for compensation requires an assessment of the probable long term impact of an electric line on private land, then an expert will give an opinion as to what would or is likely to happen. A witness as to fact can only testify as to what did (or if intentions are relevant) what was intended to happen. As explained in Appendix 23,[167] except in special circumstances a party is only allowed to call one expert witness unless the Lands Tribunal gives permission.[168] An application for permission to call more than one expert witness if made *before* the main hearing is made as an interlocutory application in the normal way.[169]

16.21.2 Duty of the expert witness

Practitioners familiar with the Civil Procedure Rules will know that the expert's duty to the court is set out in detail. The same principle is

---

165 1996R42(1).
166 PD16.1.
167 Section A.23.9.
168 This is a restriction imposed by section 2(3) of the Land Compensation Act 1961.
169 1996R42(4).

enshrined in the practice directions.[170] The expert's duty is to assist the Lands Tribunal on matters within his/her expertise. It is not the duty of the expert when giving expert evidence to act as the paid advocate of the party calling him. The Lands Tribunal expects the evidence to be accurate and complete as to relevant facts and should represent the honest and objective opinion of the witness. This may not necessarily be the same as the argument (expressed as an opinion) which best helps the party calling the expert. Obviously an opinion based on ill-considered argument or assumption is liable to be exposed under cross-examination.

However, even apart from any wish to avoid such a situation the duty to the Lands Tribunal has the potential to conflict with the expert's duty to the client by whom the expert is paid. For this reason most professional bodies adopt codes of practice and professional conduct to deal with any such potential conflict. The Lands Tribunal expects any expert witness who is a member of any such professional body to comply with any relevant code in the preparation and presentation of his evidence.[171]

16.21.3 Where more than one party intends to call expert evidence

If made at the hearing the application is made to the Lands Tribunal itself. However, such a late application is very likely to expose the applicant to an adverse costs order and possibly to the risk of an adjournment on terms. Wherever possible such applications ought to be avoided.[172]

The provisions of 1996R42(5) state that a party shall within 28 days of receiving a request from the registrar send to him and to the other parties to the proceedings a copy of each of the documents listed in that rule relating to the evidence to be given by each expert witness. That would suggest that each party prepares his expert evidence in isolation from the other party to the reference as a unilateral exercise. That is not in fact how it is supposed to work in practice. Where more than one party is intending to call expert evidence in the same discipline the experts are expected to co-operate in the preparation of

---

170 PD16.2.
171 PD16.2.
172 See 1996R42(6).

their evidence before exchanging it. The purpose of this practice direction is to ensure as far as possible that the experts are basing their opinions against a background of agreed facts.

This would include land descriptions and measurements. The practice direction also specifically mentions comparable transactions on which an expert propose to rely, as well as plans, documents and photographs. The advantage of this approach is that it limits the scope for the expert evidence to be unhelpful to the Lands Tribunal due to a reliance on assumed facts which are not in the result proved to the satisfaction of the Lands Tribunal. Of course, it is not always possible to agree everything before exchanging experts reports and so where there are areas of disagreement these must be noted. They will obviously be explored during the oral evidence given at the hearing.

16.21.4 Form and content of the expert's report

In addition to the categories of documents listed in 1996R42(5), PD16.4 sets out further specific requirements regarding the presentation of reports and their contents. These are designed to enable the Lands Tribunal to evaluate the authority and credibility of the report and the limits of the personal knowledge of the author. These stipulations will require careful consideration when the expert is a member of a large professional firm and draws upon the collective knowledge of his/her organisation. This most notably occurs when the firm in question is divided into numerous specialist departments.

The requirements are that the expert's report should be addressed to the Lands Tribunal and not to the party from whom the expert has received his/her instructions. It should:

- give details of the expert's qualifications
- give details of any literature or other material on which the expert has relied in making the report
- say what, if any, inspections or investigations the expert has used for the report and whether or not the inspections or investigations have been carried out under the expert's supervision
- give the qualifications of the person who carried out any such inspection or investigations and
- where there is a range of opinion on the matter within the report:
  (a) summarise the range of opinion and
  (b) give reasons for his/her own opinion

- contain a summary of the conclusions reached
- contain a statement that the expert understands his/her duty to the Lands Tribunal and has complied with that duty and
- contain a statement setting out the substance of all material instructions (whether written or oral). The statement should summarise the facts and instructions given to the expert which are material to the opinions expressed in the report or upon which those opinions are based.

The provisions of the practice directions[173] then contain an equivalent of something which appears in the new Civil Procedure Rules. Ordinarily instructions given to an expert witness by a party would be privileged against disclosure. That is to say the opposite party would have no right to see these documents. The original principle behind privilege was that it would enable a litigant to discuss matters with the utmost frankness with its advisers without feeling that inconvenient or damaging omissions would come to the attention of the opposite party. However, since the introduction of the Civil Procedure Rules, there has been a change of approach. Instructions to an expert witness are not privileged against disclosure, but it is not the ordinary rule that the opposite party will automatically be able to see those instructions. The Lands Tribunal will not, in relation to instructions to an expert, order the disclosure of any specific document or permit any questioning in the Tribunal other than by the party who instructed the expert, unless it is satisfied that there are reasonable grounds to consider the statement of instructions given to the expert to be inaccurate or incomplete. It follows that if there is going to be any attempt to investigate the substance of such instructions, there has to be a case which the Lands Tribunal will regard as reasonable before it will order disclosure or permit questioning. Mere fishing expeditions to see whether there is anything potentially useful in the instructions will not be allowed.

An expert's report should be verified by a statement of truth, as well as statements required by the practice direction as cited above. Members of the Royal Institution of Chartered Surveyors should comply with the form of declaration contained in "surveyors acting as expert witnesses — practice statement" issued by that institution. The form of statement of truth is "I believe that the facts stated in this witness statement are true and that the opinions expressed are correct".[174]

---

173 PD16.5.
174 PD15.1.

16.21.5 Lodging reports

The consistent theme of all the practice directions is that the parties should be able to prepare with the utmost efficiency for any hearing, avoid wasting time at the hearing and not be taken by surprise by any of the evidence which comes out at the hearing. For this reason, directions given by the Lands Tribunal will normally require that experts' reports are lodged with it and exchanged with the other side at an early stage prior to the hearing. The normal rule is that an expert can only give evidence at the hearing in respect of matters disclosed in the expert's report. If it is thought desirable that each expert should respond to a report of another served in the same reference, then a further report has to be prepared and served in good time and then that is treated as notice of intention to give additional evidence for which formal permission of the Lands Tribunal must be sought at the hearing, if not before. Expert's reports must not contain any reference to or details of without prejudice negotiations or offers of settlement.[175]

There is a significant provision in the practice directions but it appears not under lodging of experts reports but under the heading "written questions to experts" which we consider below. However, it is important on the question of what evidence can be used at any hearing. The practice direction states that where a party has disclosed an expert's report, any party may use that expert's report as evidence at the hearing.[176] Consequently, if you want to use, for instance, comparable evidence or other research contained in the opposite party's report positively to support your own case, then you are at liberty to do so.

16.21.6 Written questions to experts

In relation to a report disclosed by an opposite party in a reference, the other party has the right to put written questions to the opposite party's expert. It is quite wrong to send questions to the expert directly without copying them to any solicitors who are on the record. Normally there should be one single set of questions put once only and they should be put within 28 days of service of the expert's report and should be put only for the purposes of clarification of the report.[177]

---

175 PD16.7.
176 PD16.10.
177 PD16.8.

To make the best use of this opportunity it is important that the questions are first debated and refined not only between the party and his own expert witness, but also the party and his legal advisors. In particular, if counsel is going to be doing the advocacy at the hearing, then the best practice is to ensure that as part of your own private timetable for the preparation of the reference for hearing, you have allowed for a conference with your counsel immediately after the date on which expert reports are due (allowing for any agreed extensions) so that the discussion on your side can occur at the beginning of the 28 period rather than towards the end. It should be noted that any questioning which does not fall within the parameters of the practice direction is highly likely to result in an interlocutory application to the Lands Tribunal for a ruling on whether the question is permissible.

The practice direction goes on to say that the costs incurred in dealing with any written questions are payable in the first instance by the party asking them. This does not stop the Lands Tribunal at a later date making an order for costs in favour of the investigating party which results in a reimbursement of any such charges. However, the fact that such charges will be incurred and payable in the first instance is another reason for ensuring that the opportunity is used carefully and efficiently.

An expert's answer to any request is put in writing and will be treated as part of that expert's report.

It is of course, possible that the answers received will not be as full or as helpful as the party asking them would have wished. There may, in certain circumstances, be an explicit refusal to answer the question for any reason. In those circumstances the matter must be referred back to the Lands Tribunal or the registrar on an interlocutory application for an order directing an answer to the question. If the questions are answered, the Lands Tribunal and the registrar may make one or both of the following orders in relation to the party who instructed the expert who is refusing to answer:

- that the party may not rely on the evidence of that expert or
- that the party may not recover the fees and expenses of that expert from any other party.

As a matter of best practice, given that there will invariably be a without prejudice meeting between the experts (see below) the ability to answer the questions would actually be better exercised after issues have been clarified at that without prejudice meeting. This obviously can cause a degree of difficulty with the timetable, bearing in mind that normally written questions to experts must be put within 28 days of service of the report and that would leave very little time after any without prejudice meeting between them. Accordingly, it is worth considering this point at the very outset of the making of any directions so that an appropriate timetable can be agreed between the parties and approved by the Lands Tribunal.

16.21.7 Discussions between experts

The principle of without prejudice discussions between experts is now well established under the Civil Procedure Rules and an equivalent provision is now found in the practice directions.[178] The meeting will usually be ordered as part of the directions made by the Lands Tribunal. It will take place after the exchange of expert reports. The meeting is held on a "without prejudice basis" between experts of a like discipline so that the experts can debate the issues freely without fearing that they will compromise their clients by an inappropriate admission. The aim of such a meeting is to provide an opportunity for the experts to reach such agreements as they can on matters of fact, plans, photographs, measurements, comparables and similar points with a view to narrowing the issues for the final hearing. The policy behind the provision for this kind of meeting is to encourage out of court settlement and limit the time spent in cross-examination of the experts at the final hearing and so minimise the costs incurred.

The practice direction specifically states that the discussions between the experts are not to be mentioned at the hearing unless the parties agree. Further an agreement reached at the meeting on an issue is not to be binding on the parties unless they expressly so agree.

This then raises a number of questions as to how such meetings should be conducted and recorded. It is invariably unproductive if the experts approach the meeting with instructions to make no concessions. It is also unrealistic to expect that some negotiation of the claim will not occur even if formal offers are not made. Nevertheless, the experts need to be careful that they do not compromise themselves in terms of their duty to the court. It is far better for the experts to confine themselves to the matters within their discipline and allow themselves to be challenged by the analysis of their opposite number.

---

178   PD16.

In the long run that may be much more beneficial to the conduct of the case in that misunderstandings and weaknesses are identified before the main hearing rather than in cross-examination.

It would be inappropriate and counter-productive for the meeting to have lawyers in attendance as note-takers. The outcome of any without prejudice meeting of experts is not supposed to be a Hansard-like record of the discussion. It is supposed a statement of the matters agreed and the matters not agreed and the reason why. Again it is inappropriate for much time to be spent on the drafting of such a statement as if each particular word has to be examined for hidden meaning. It is the record of the extent of agreement and disagreement which is vital.

As a practical measure, we recommend that the parties exchange open letters to the effect that nothing will be treated as expressly agreed for the purposes of using information discussed at the meeting at the hearing unless a statement jointly signed by the experts has been served by one party on the other and lodged with the Lands Tribunal.

This avoids introducing any subsidiary dispute into the reference.

16.21.8 Computer-based valuations

Where valuers propose to rely upon computer-based valuations they must agree upon a model which can be made available for use by the Lands Tribunal or otherwise the parties must in default of agreement seek directions.[179] Any such application will be an interlocutory application made in the usual way.[180]

16.22 Site inspections

The usual practice will be that an expert preparing a report for the purposes of a reference to the Lands Tribunal will have visited the land in question at least once, and indeed may even have done so in the company of the opposite expert. Their reports obviously will be informed by the impressions gained on that inspection. It can also be the case that the Lands Tribunal itself will, in an appropriate case benefit from the visiting the land which is the subject of the reference

---

179 PD16.13.
180 1996R38 discussed in section 16.14.5 above.

and where practicable any comparables referred to by the experts.[181] An express power is given to the member of the Lands Tribunal dealing with the reference to enter upon and inspect any land which is the subject matter of the reference.[182]

Notice ought to be given by the parties who shall be entitled to be represented at the inspection. The Lands Tribunal will decide how many representatives from each party it will permit. By way of general guidance the Lands Tribunal states that normally only one representative from each side will be permitted, unless the parties otherwise agree.[183]

The practice direction makes it clear that the Lands Tribunal will not accept any oral or written evidence tendered in the course of the inspection. This can create a practical difficulty. If the purpose of the inspection is to point out matters which have been mentioned in an expert's report, then inevitably one representative on either side must speak to the members of the Lands Tribunal who are making the inspection. In those circumstances there ought to be a second person present to make notes. The best practice would be that the outcome of any inspection be recorded in a jointly signed statement of facts agreed and not agreed and/or any agreed annotated plan.

The first question always, however, to be asked, is whether the Lands Tribunal would not be equally well assisted by photographs. Photographs can be misleading, of course, but care needs to be taken that an inspection yields the information which the Lands Tribunal most needs without prejudicing either party.

The Lands Tribunal may also make an unaccompanied inspection without entering onto private land.

16.23 Assessors

The President of the Lands Tribunal does have power in respect of any case coming before the Lands Tribunal to direct that the Lands Tribunal shall hear the case with the aid of one or more assessors appointed by him whose remuneration shall be determined by the President with the approval of the Treasury.[184] The point about the

---

181 1996R29.
182 See section 2(2) of the Land Compensation Act 1961.
183 PD19.
184 1996R29A.

assessor is that the assessor will have specialist knowledge not available to the President or other members of the Lands Tribunal. Bearing in mind that the members of the Lands Tribunal will usually be lawyers or surveyors, the inference to be drawn is that some other specialist expert discipline is needed to inform the discussions of the Lands Tribunal once the evidence is heard. This means that the assessor will need to be present during the whole length of the hearing so that his contribution is reflective of and responsive to that evidence.

The Lands Tribunal will say that it must appear to the President that an assessor is required. It does not go on to say that either party can so apply. However, whether or not an assessor is to be appointed ought to be a matter considered by the parties and raised directly with the Lands Tribunal at any directions hearing or pre-trial review.

16.24 Negotiations, settlements and withdrawals

References will very often be concluded by a negotiated settlement rather than a hearing and a decision of the Lands Tribunal. Indeed, references to a Lands Tribunal naturally lend themselves to settlement negotiations. The issues between the parties will be expressed in the rival amounts of compensation said to be payable and thus there will be a relatively clear understanding of the amount at stake. The costs of pursuing the matter can therefore be reasonably estimated and then a costs benefit analysis undertaken. The case is usually highly dependent upon expert opinion, the strength of which can be tested at the without prejudice meeting between the expert witnesses to which we refer above.

The Lands Tribunal follows the current practice of officially encouraging settlement discussions. Accordingly, when the Respondent replies to a notice of reference (see section 16.14.3) the Respondent can ask the Lands Tribunal to stay the proceedings for four weeks to enable the parties to engage in alternative dispute resolution. If the claimant agrees then the stay will be ordered.[185] A longer period will be considered, but only if the parties can satisfy the Lands Tribunal that the circumstances justify it. Alternative dispute resolution can mean any form of resolving the problem on which the parties agree and will

---

185 PD4.

usually encompass mediation, third party expert determination or an informal arbitration.

It may be that the parties are not able to use this opportunity so early in the reference. That does not mean that settlement negotiations cannot continue at any time during the course of the reference. However, the Lands Tribunal makes it clear that the fact that negotiations are continuing will not usually be taken as the sort of exceptional circumstance that justify the postponement of a hearing.[186] If without prejudice negotiations are likely to occur, then the best way of accommodating space for those to take place is to be relatively generous in the directions timetable which is agreed with the other side and ordered by the Lands Tribunal.

Those familiar with the Civil Procedure Rules and rent review negotiations will be aware of the concept of a Part 36 offer/payment and/or a Calderbank offer. The principle behind such offers is that if the party receiving the offer fails to obtain following the final hearing before the Lands Tribunal a sum in excess of, or a result better than that contained in, the offer, then (even though the recipient of the offer may have "won") that party is at risk of being penalised in costs. Thus the benefit of any win may well be reduced if the recipient of the offer is going to be exposed for a significant proportion of the costs of achieving it.

The Lands Tribunal has an equivalent procedure in the form of sealed offers.[187] Whereas under the Civil Procedure Rules when legal action is underway, an amount must actually be paid into court if a money offer is going to made, before the Lands Tribunal all that is necessary is that an offer be made and put in a sealed envelope and then the sealed envelope is in turn lodged with the Registrar of the Lands Tribunal. The Lands Tribunal is not allowed to consider the contents of the sealed offer until after it has made its decision on the merits of the case. The sealed offer is then opened when that decision has been published. The sealed offer cannot be used to affect the Lands Tribunal's decision on the amount of any compensation payable. It will, however, be used when the Lands Tribunal comes to decide who should pay the costs of the reference and in what proportions.

---

186 PD13.
187 1996R44. Note this is different from the unconditional offer regime in section 4 of the Land Compensation Act 1961. A sealed offer is made on a without prejudice basis and is not one which can be mentioned to the Lands Tribunal until after the award.

The sealed offer, of course, is made directly to the opposite party. If they should accept the offer or even without a formal sealed offer being made, negotiations result in a settlement, then the reference may be withdrawn by sending to the registrar a written notice of withdrawal signed by all parties to the reference or their representatives.[188] As a matter of best practice, the Lands Tribunal needs to be advised at the earliest possible opportunity that any fixed hearing date will no longer be required. Any party waiting until the actual day of the hearing to notify the Lands Tribunal of a prior settlement will be at risk of having to pay the wasted costs of the day, regardless of the outcome of the settlement.

16.25 Consent orders

If the outcome of the settlement negotiation is that an agreed sum of compensation is paid by the licence-holder to the claimant in recognition of the grant of a necessary wayleave or of disturbance to movables or to the right of enjoyment of land or movables, then it may not be necessary to have any further form of order from the court. A mere withdrawal will be sufficient, particularly if the sum is paid promptly and well ahead of any fixed hearing date.

However, it may be the case that something needs to be done as a term of the settlement; for instance: a document to be executed. Alternatively, there may not be enough time before any fixed hearing date to arrange a payment of the money and so the claimant will want an enforceable obligation for the money to paid within so many days after the negotiation of the settlement. Those are the kinds of cases where a consent order may be required.

A consent order is an order of the Lands Tribunal. The only difference is instead of the Lands Tribunal imposing upon the parties its own view of matters, the parties themselves agree to the terms of the order and any specific details relating to the completion of the settlement are usually set out in a schedule to the order. The model which such consent order normally follow is that of a Tomlin Order. There is some very useful guidance about the drafting of such orders in CPR40.6 and the editorial comment upon that rule in 40.6.2 of the Supreme Court Practice. The schedule to any consent order acts as a contract known

---

188   1996R45.

among solicitors as a contract of compromise. It must therefore be drafted with the same degree of precision as any other contract would demand. Indeed in recent times there has been a growth in the number of claims brought against solicitors for the negligent drafting of consent orders.

Once the terms of a consent order have been settled, whatever they may be, the parties should ensure that they have all signed a single document containing all of the terms and that document is then sent to the Registrar who will arrange for the order to be made.[189]

16.26 Sanctions

We have already discussed above the steps which the Lands Tribunal can take if it considers that there are deficiencies in the presentation of the expert evidence. It may, in other circumstances, be necessary for the Lands Tribunal to impose sanctions. This would, for example, be relevant if one party consistently fails to obey the directions of the Lands Tribunal. Any party wishing to invite the Lands Tribunal to impose sanctions must make a normal interlocutory application.

The most extreme sanction is a dismissal of the reference without a hearing on the merits. A party may, at any time, apply for the dismissal of the proceedings if he thinks it is appropriate so to do. The application however must be made direct to the President and not the registrar.[190]

An application for dismissal may also be considered if it is thought that the reference has been wrongly commenced or commenced on an incorrect notice. Whatever may be the circumstances, it is clearly a very serious application to make and one only to be considered maturely before being launched.

There may be other instances where a sanction is needed in order to move the reference along in respect of which dismissal would be an inappropriate and disproportionate response. In those circumstances the Lands Tribunal has power to order the service of documents if that is required, or to require any party at fault to pay any additional costs occasionally as a result of the failure or if necessary adjourn the hearing of the proceedings until the breach has been remedied.[191] If

---

189   1996R51.
190   1996R45.
191   1996R46(1).

there is evidence of a failure to pursue any reference with due diligence then the Lands Tribunal likewise, provided it first hears the party alleged to be in default, may make an order fixing the trial date, notwithstanding any apparent lack of preparation, or may make an order dismissing the proceedings or debarring the defaulting party from taking further part in the proceedings. Alternatively the Lands Tribunal may make any other order appropriate for expediting the proceedings or disposing of them, including an order for costs.[192] Any of the above applications needs to be made as a normal interlocutory application to the registrar.

It should be noted that any failure by any person to comply with the 1996 Rules does not render the proceedings or anything done in pursuance of them invalid unless the President or the Lands Tribunal so directs.[193]

16.27 Appeals

Section 3(4) of the 1949 Act provides that the decision of the Lands Tribunal is final. However, the proviso to that section does permit an appeal on a point of law. Originally the 1949 Act required the Tribunal to state a case for the benefit of the Court to which the appeal was directed. This is no longer the procedure. An appeal from the Lands Tribunal is made directly to the Court of Appeal.[194] In relation to that Appeal, the provisions of Civil Procedure Rule 52 must be followed. This means that first permission to appeal is required from the Court of Appeal. This is requested in the form of the appellant's notice which the appellant must file at the Court of Appeal within 28 days after the date of the decision of the Tribunal.

The decision of the Tribunal takes effect for this purpose on the day on which it was given unless the decision states otherwise. Usually the decision will state that it will take effect when and not before the issue of costs has been determined. Bearing in mind that an appeal to the Court of Appeal lies only on a question of law, permission to appeal will only be granted if the Court of Appeal considers that there is an arguable case that the Lands Tribunal has misdirected itself on a point of law.

---

192 1996R46(2).
193 1996R47.
194 Civil Procedure (Modification of Enactment) Order 2000.

16.28 References under 1996 Rules Part VII: application of the Arbitration Act 1996

The 1949 Act (as subsequently updated) does allow the Lands Tribunal to act as an arbitrator under the Arbitration Act 1996. Rules imposed by the Lands Tribunal upon itself can limit the extent to which it exercises the arbitral jurisdiction.[195]

References by consent in respect of which the Lands Tribunal is acting as an arbitrator are dealt with by 1996 Rules Part VII. 1996R26 sets out the particular provisions of the Arbitration Act 1996 which apply to the reference unless the parties otherwise agree in addition to those set out in 1996R32.

Two points require further comment here. First, why would parties appoint refer a matter to the Lands Tribunal for such a determination? It is the essential feature of arbitration that the parties are in control of the procedure rather than the arbitrator, although the arbitrator does have a statutory duty to act fairly between the parties. It is therefore intended to be a more flexible and informal method of dispute resolution.

Second, 1996R32 applies to all forms of references to the Lands Tribunal as part of the general procedure, aspects of the Arbitration Act which are considered beneficial to the expeditious disposal of those proceedings. As an apparent derogation from the principle of party autonomy, 1996R32 appears to be an non-negotiable aspect of any arbitration procedure. However that would, in our view, be a perverse reading of 1996 Rules Part VII. We consider that the proper reading of 1996R26 is to treat "unless otherwise agreed" as applicable to both 1996R26 and 1996R32.

Subject to those points the parties may incorporate as the rules of their arbitration any other rules on which they agree. Many established arbitration agencies or courts of arbitration will have standardised rules which are applied to disputes coming before them. Model rules may be adopted if appropriate. Alternatively the parties may decide that they will adopt some or all aspects of the general procedure discussed above.

The only other point to bear in mind is that when notice of reference is given, the Lands Tribunal must be supplied with copies of the arbitration agreement.[196] In this connection the reader is referred to the discussion in section 16.7.

---

195 Sections 3(6)(c) and 3(8) of the 1949 Act.
196 1996R26A.

Part 5

Enforcement of Rights

Relevant Remedies

17.1 Introduction

Inevitably, with the best will in the world, circumstances will from time to time arise in which either a licence-holder or a landowner or an occupier has cause to complain that obligations have not been duly performed.

If the complaint arises in connection with a non-statutory wayleave, it could be that the grantor has denied access and inspection rights to the licence-holder or that the licence-holder has failed to make the wayleave payments in due time.

If the complaint arises in connection with a necessary wayleave which has been granted the possibilities are similar to those set out above. The financial obligation will be to pay such compensation as the Lands Tribunal has assessed whether that is by way of a lump sum or by way of a periodical payment.

If a necessary wayleave application to the Secretary of State has been refused, then the most likely complaint is that the licence-holder has failed to remove the electric line within the period prescribed by paragraph 8(4) of schedule 4 to the Act.[1] If there is no wayleave at all, and never has been and the licence-holder without involving the procedure laid down by paragraph 6(1) of schedule 4 to the Act,[2] proposes to install a new electric line, then the landowner or occupier would be concerned to prevent a trespass onto his property.

---

1 Namely one month from the date of the refusal or such longer period as the Secretary of State may specify.
2 See Chapter 9.

A possibility common to both non-statutory wayleaves and to necessary wayleaves is that the licence-holder exceeds the terms of the wayleave and so makes wrongful use of land.

Finally, with existing lines, there could be questions as to whether there is any liability to make further wayleave payments once the statutory procedures described in Chapters 9 and 10 have been instigated.

17.2 Outline of chapter

Before considering which of the available remedies are appropriate for these various circumstances, it is necessary first to say something about remedies in general. Accordingly, in this chapter we are concerned with the definition of the four most significant remedies for our own purposes:

- debt
- damages
- mesne[3] profits
- injunctions.

In subsequent chapters are considered the specific points arising on an application to any particular situation of the general principles we discuss below. Other remedies exist and may be relevant in any particular situation, but a wider discussion is outside the scope of this book.

17.3 Debt

A debt is a sum of money due and owing in return for the performance of a corresponding obligation. It is an essential precondition of a debt that it is a fixed amount which is known before legal action begins or is capable of being calculated. A debt is, in the technical language of the law, liquidated and ascertainable.

Recalling the contractual basis of wayleaves,[4] the obligation to pay the wayleave payment in return for the permission to keep the electric line creates a debt due from the licence-holder to the grantor.

---

3 Pronounced "mean" — the word is derived from Norman French.
4 See Chapter 5.

A debt by definition is due by a certain date. Consequently if it is paid after the due date it is eligible to attract interest as compensation for the delay in payment. Interest may be contractually payable if there is a term to that effect in the contract or may be payable pursuant to statute[5] or may be paid at the discretion of the court.[6] A failure to pay a debt is a breach of contract just as much as a failure to perform some other obligations for which damages are sought.

17.4 Damages

Damages is the term given to the financial compensation which a court orders one person to pay to another. The law on damages is complex and extensive and a detailed treatment is outside the scope of this book. Suffice it to say for our purposes that a distinction must be drawn between damages payable as a result of a breach of contract (contractual damages) and damages payable as a result of injury or harm caused by one person to another (damages in tort). Contractual damages are designed to be a substitute for the performance of the contract which has been broken. They are supposed (so far as money can do this) to put the injured party in the position he would have been in if the obligation had been duly performed. Damages in tort are designed to compensate for actual loss and damage suffered.

Damages are awarded by a court after an investigation into the complaint. In legal terms, proceedings are commenced and a civil trial takes place. By definition the amount of the damages is not known at the date legal action commences. They are unliquidated. The claimant must prove the breach of contract or the liability for loss and damage suffered. The claimant must also demonstrate in his evidence the proper level of compensation and any entitlement to interest. Damages are liquidated and so become a debt when the court awards the specific sum it judges to be the right amount.

---

5   Late Payment of Commercial Debts (Interest) Act 1998. This entitlement is only available to one business claiming against another.
6   Section 69 of the County Courts Act 1984 (for county court proceedings) or section 35A of the Supreme Court Act 1981 (for High Court proceedings).

17.5 Mesne profits

Mesne profits are intended to compensate a landowner or occupier for wrongful use of his land.[7] They are usually calculated on a daily basis by reference to the annual value of the land used and are payable for the length of time during which the wrongful use continued. However, whichever method of assessment is used, the aim is to identify the fair value of the use to which the trespasser has wrongfully put the land. The land used by an electric line would extend to the conductors over the land as well as the overhead line supports. The payments referred to in Chapter 14 relate to recompense for the use of land by licence-holders on agreed terms and would not necessarily represent the same basis of calculation for the wrongful use of land by licence-holders. However having regard to the analogy with rent discussed in Chapter 14, and the decision in *Clifton Securities Ltd v Huntley*,[8] they would be an obvious starting point. A Claimant may need to elect whether the mesne profits claim is in substitution for a damages claim or whether a damages claim for actual injury can be made alongside it.

17.6 Injunctions

The technical term for a court order to do or not to do something is an injunction. If it is an injunction requiring an act to be done it is mandatory. If it is an injunction restraining or preventing a wrongful act from being done it is prohibitory.

So far as injunctions are concerned, for the purposes of this book, we cannot go into the detail which would be found in a specialist litigation text book or a commentary on the Civil Procedure Rules. A treatment in outline must suffice.

The overriding objective of an injunction is to provide a remedy where financial compensation would not be adequate. Injunctions are granted if the court thinks fit on the evidence presented to it. There is no automatic right to an injunction. They are granted at the court's discretion. Nevertheless the court has well established principles to which it refers when deciding whether or not to grant one.

---

7   A full discussion of the relevant law can be found in *Unlawful Interference with Land* by David Elvin QC and Jonathan Karas QC.

8   [1948] 2 All ER 283.

For instance, if either before or just shortly after the commencement of an unjustified intrusion onto land, a landowner became aware that such an intrusion was planned or had just begun and such landowner complained that he had not consented, he could begin legal action. He issues a claim form and embarks on the litigation process which leads to a civil trial.

However, an owner or occupier of land will say that if the intrusion continues regardless while the litigation process unfolds, a judgment after trial will be of no practical benefit. The damage will be done by then. So an owner or occupier of land would ordinarily ask the trespasser to refrain from intruding pending the outcome of the legal action. If the trespasser refuses or ignores the request, an owner or occupier of land asks the court for emergency assistance. He seeks an order to prevent the intrusion being continued pending trial. This is called an interim injunction.

In such case, the questions the court asks itself are:[9]

- Is there "a serious question" to be tried?
- Would damages be an adequate remedy?
- If not, does the balance of convenience favour the grant of an injunction?
- Are there any special factors weighing against the grant of an injunction? In particular has the claimant unreasonably delayed the making of the application?

Bear in mind that at this stage the court has not examined in detail the rights and wrongs of the matter. It has taken a view on limited information as to what needs to be done in the particular situation to preserve the status quo pending trial. Hence it asks any claimant who seeks an interim injunction to give to the court an undertaking that if at trial it should turn out that the interim injunction ought not to have been granted, then the claimant must compensate the alleged trespasser who obeyed the injunction for any losses suffered as a result of so obeying. The giving of such an undertaking is a necessary precondition for the grant of an interim injunction.

When it comes to trial, the court has a further choice to make. If having examined all the evidence it concludes that the claimant is in the right, it can grant a permanent injunction which either rectifies any wrong

---

9   As laid down in the leading case of *American Cyanamid Co (No 1)* v *Ethicon* [1975] AC 396.

already done (ie it is a mandatory injunction) or it prevents any further wrong doing from being repeated (ie it is a prohibitory injunction).

However, the court does not have to do this. Section 50 of the Supreme Court Act 1981 allows a court in any case where it could grant an injunction to award damages in addition to or in substitution for such an order.

If damages are awarded in addition to an injunction then not only is the wrongful act banned permanently for the future, but also compensation is paid for any physical injury to the land or for any wrongful use of the land prior to trial.

If damages are awarded in substitution for an injunction then as well as dealing with loss and damage suffered prior to trial the damages also act as compensation for such future wrongful activity as will occur after the trial.[10]

The question then arises in what circumstances will the Court exercise this power? A working rule was laid down by the Court in *Shelfer v City of London Electric Lighting Company* [1895] 1 CH 287 which is still applied today. In essence if the wrong done to the claimant is comparatively small and can be adequately compensated by an award of damages and if by contrast a grant of an injunction would be oppressive to the defendant, a permanent injunction will be refused. It is an interesting coincidence that the case should include an authorised electricity undertaker prior to the 1947 Act!!

As an example from more recent times of the court using that power, a builder was allowed to retain houses built in breach of a restrictive covenant provided 25% of the development profit was paid over to the aggrieved landowner.[11] The court has also used the power to "licence" a house built in a cul-de-sac in the mistaken belief that the access to it was a public highway when it was in fact a private roadway protected by restrictive covenants over which the occupants of the new house would inevitably trespass every time they went to and fro.[12]

---

10 *Leeds Industrial Co-Operative Society v Slack* [1924] AC1 851 decided under the Chancery Procedure Amendment Act 1858 ("Lord Cairn's Act") the statutory predecessor of section 50 Supreme Court Act 1981 held that such was the scope of the power given to the Court.
11 *Wrotham Park Estate Co Ltd v Parkside Homes Ltd* [1974] 1 WLR 798.
12 *Jaggard v Sawyer* [1995] 1 WLR 269 The Court upheld the view that an injunction would be oppressive. It upheld a valuation of £6,250 as the fee by which the restrictive covenant could be bought off.

17.7 Limitation of actions

Finally, it should be noted that the right to enforce any obligation by legal action or to sue for wrong done can be permanently lost. A detailed examination of the law of limitation is outside the scope of this book. Suffice it to say as a working rule in the context of wayleaves, that legal action for breach of contract must be brought within six years of the date of the breach and for damages in tort, six years from the date of which the harm or injury was suffered.

Failure to Comply with Financial Obligations

18.1 Outline of chapter

There are two financial obligations under consideration in this chapter. The first is the obligation to make a wayleave payment due under the terms of an express or an implied wayleave. The second is the obligation to pay any compensation awarded by the Lands Tribunal following the grant of a necessary wayleave or as a consequence of the exercise of statutory rights. In both cases it is for the licence-holder to perform the obligation. In both cases if the obligation is not performed, it is the landowner or occupier who has the right to seek a remedy for that failure. This chapter discusses in outline how a remedy for each such failure is obtained. Before considering the remedies available to deal with each such failure it is important to review how the relevant obligations actually come into existence. It is important not to assume that such obligations exist. The successful pursuit of any of the remedies considered below presupposes that the existence of the obligation in the first place can be demonstrated. It cannot be simply inferred from the fact that an electric line is present on the land.

18.2 Sources of obligation to pay: wayleave payments

A wayleave payment is a debt due from the licence-holder to the landowner or occupier (as the case may be) who granted the wayleave ("the grantor").

Where payments are made under a non-statutory wayleave, provision for payment is provided within the contract (if express) or will follow the established course of dealings (if implied). Where that contract is brought to an end then there is no continuing provision for payment. This creates an anomalous situation where the existence of the electric line continues to impact upon a landowner, but the provisions under schedule 4 to the Act for compensation to be paid have not yet been brought into play. Such a situation arises in the period from the date of termination of the non-statutory wayleave, to the date on which a necessary wayleave is granted by the Secretary of State.

While it may be the policy of some licence-holders to continue to offer the previous rates of payment over that period as a purely gratuitous goodwill gesture, there is no statutory obligation for them to do so. Furthermore, the grantor may justifiably be reluctant to accept such payments in case it gives the licence-holder an argument that a fresh implied wayleave has been granted so that the whole two notice process[1] has to begin again. This is a problem we consider further in Chapter 22.

Care also needs to be taken if there has been an attempt to review the levels of wayleave payment under a non-statutory wayleave. Unless such a review has taken place strictly in accordance with the applicable contractual provisions, it may be the case that the original payment obligation has been inadvertently terminated and some other more limited or just different payment obligation has been agreed between the parties in substitution for the original.

18.3 Sources of obligation to pay: compensation

Paragraph 7 of schedule 4 to the Act provides that where a wayleave is granted to a licence-holder both the occupier of the land and the owner may recover from the licence-holder compensation in respect of the grant.

Additionally paragraph 7 provides that a claim for compensation may be made if there is damage to land or moveables or disturbance in the enjoyment of land or moveables.

---

1   See Chapter 9 and specifically section 9.6.

18.4 Statutory demand

Contractual debts can be collected in two ways.

Where there is an undisputed debt in excess of £750 a demand in the prescribed form[2] can be served upon a limited liability company at its registered office. Licence-holders are all limited liability companies. Registered offices can be obtained by visiting the Companies House web-site[3] and inserting the name of the licence-holder on the page marked select and access company information. When the name of the licence-holder is displayed, click on the corporate number.

The effect of the service of a statutory demand is that if the licence-holder does not pay the sum due within 21 days of service, the licence-holder is deemed to be insolvent provided the licence-holder is genuinely indebted to the landowner or occupier. A winding-up petition can be presented after the 21 days have elapsed unless the licence-holder has obtained an injunction to restrain the presentation of a petition. A detailed discussion of insolvency procedure is outside the scope of this book. Suffice it to say that invariably the service of a statutory demand and, all the more so, a winding up petition will prompt immediate payment except where the debt is disputed. The courts have decided that even if a company is not in actual fact insolvent, the deeming provision takes precedence.[4]

18.5 Legal action through the county court

If a debt due does not exceed £750 or if the debt (of whatever size) is disputed, the proper way to recover the debt is by legal action through the county court.

Before issuing proceedings it is the accepted practice for a grantor (or the grantor's advisers) to write to the licence-holder demanding payment. The letter should contain the following elements:

- A statement of authority to act — if not written by the grantor personally

---

2   Form 4.1 prescribed by Rule 4.5 of the Insolvency Rules 1986 (as amended) and sections 123(1)(a) of the Insolvency Act 1986.
3   *www.companieshouse.gov.uk*.
4   *Cornhill Insurance plc* v *Improvement Services Ltd* [1986] 1 WLR 114.

- a statement as to whether the non-statutory wayleave is voluntary or implied. If voluntary and express oral — recite the terms agreed; if voluntary and express written,[5] enclose a copy of the signed agreement and refer to the terms. If implied describe how the wayleave has been implied from conduct and circumstances
- a statement of the amount alleged to be due and how the amount has been calculated;
- a deadline by which payment must be made.

If payment is not made, the legal action may be commenced in the county court and pursued in accordance with the Civil Procedure Rules,[6] a further discussion of which is outside the scope of this book.

18.6 Interest

Interest may be payable on overdue debts in any of the three circumstances mentioned in Chapter 17 section 17.3. However, except in the case of contractual interest or claims to which the Late Payment of Commercial Debts (Interest) Act 1998 relate, you will not be able to claim interest when serving a statutory demand.

The ability to claim interest under either section 69 County Courts Act 1984 or section 35A Supreme Court Act 1981 only arises if court proceedings are issued and even then is within the discretion of the court.

Where a licence-holder exercises its powers of compulsory acquisition under schedule 3 to the Act, the requirement to pay interest on any outstanding compensation is governed by statutory instrument[7] and is discussed further in Chapter 15.

Interest arising in connection with compensation awarded by a Tribunal under schedule 4 to the Act in respect of a necessary wayleave needs to be considered in two stages. First, interest forming part of the award itself. Second, interest arising once the award has

---

5 See Chapters 6 and 7 for a discussion of the distinctions between the categories of wayleave here mentioned.
6 The Civil Procedure Rules are made under the authority of the Civil Procedure Act 1997.
7 The Acquisition of Land (Rate of Interest after Entry) Regulations 1995 SI 1995/2262.

become a judgment debt for the purposes of enforcement. This is discussed separately below. Returning to interest as part of the award, there is no explicit direct statutory authority enabling a person over whose land a necessary wayleave has been granted to claim interest as a head of loss. The Acquisition of Land (Rate of Interest after Entry) Regulations 1995 do not strictly apply to necessary wayleaves because they do not grant an interest in land. Nevertheless, the practice of licence-holders has been to apply these regulations by analogy when making offers to settle.

18.7 Failure to pay compensation awarded

The commentary in this paragraph applies equally to compensation ordered to be paid as a lump sum and to compensation ordered to be paid by way of periodical payments. If there is a failure to pay compensation awarded by the Lands Tribunal, the first step for the aggrieved landowner or occupier is to convert the Lands Tribunal award into a judgment debt. The procedure for so doing is set out in Part 70.5 of the Civil Procedure Rules 1998[8] and the practice direction mentioned there. Once the procedure has been completed the award is as fully enforceable as a judgment of the court.

Once the award has become a judgment debt it will attract interest under section 17 of the Judgments Act 1838. Interest runs from the date on which judgment is entered until payment of the sum due. The rate of interest is varied by statutory instrument. Under the Judgment Debts (Rate of Interest) Order 1993 the rate for judgments entered after 01 April 1993 is 8% pa.

---

8   Made under the Civil Procedure Act 1997. The rules are regularly updated by statutory instrument. The latest edition with or without detailed commentary can be obtained from a variety of sources including the website of the Ministry of Justice (*www.justice.gov.uk/guidance*).

Refusal to Allow Exercise of Wayleave Rights

19.1 Outline of chapter

As discussed in Chapters 5–7 inclusive, the term wayleave rights includes both the basic permission to locate and retain on land an electric line and also rights ancillary to that basic permission. If during the currency of either a non-statutory or a necessary wayleave the grantor purported to revoke the basic permission and then act upon that revocation *without* pursuing the statutory procedures considered in Chapter 9, every factor points towards an application for an injunction on the part of the licence-holder. We discuss these factors in Chapter 17. Accordingly, the remainder of this chapter is concerned only with ancillary rights.

19.2 Relevance of ancillary rights

As we have seen, the terms of a wayleave will usually be supplemented by rights of access to enable the licence-holder to inspect and maintain the electric line. These rights are often called ancillary rights and that is the description we shall use for the remainder of this chapter. Such rights are needed if the basic permission granted by the wayleave is to be exercised efficiently by the licence-holder.

19.3 How are ancillary rights granted

In the case of necessary wayleaves, the court has decided that ancillary rights must always be implied into the wayleave.[1]

In the case of express non-statutory wayleaves, ancillary rights will only exist if they are given by the terms of the documents signed by the current landowner/or occupier as grantor and the current licence-holder. If they are given by that document, they will usually be the subject-matter of an express provision. Occasionally, however, they may be implied into a voluntary wayleave on the basis of the apparent intention of the parties judged by the wording of the voluntary wayleave. Such an implication can be justified if it is necessary in order to give the wayleave business efficacy.[2]

In the case of an implied non-statutory wayleave, ancillary rights will only be granted if the circumstances from which a wayleave is presumed to have arisen, likewise justify the deduction that such rights have also been granted.

19.4 Legal questions arising from a refusal

Assuming that ancillary rights have been granted by one means or another, the question is what remedy is available to the licence-holder if a landowner or occupier refuses to permit the licence-holder to exercise them? The circumstances are realistically only likely to arise if a non-statutory wayleave has lapsed or automatically expired[3] and the current landowner or occupier is objecting to the exercise of ancillary rights but has not served a notice to remove the electric line.

As explained in other chapters,[4] the landowner or occupier, cannot even following an automatic lapsation or termination of a non-statutory wayleave, require the licence-holder to remove the electric line until first a notice to remove has been served and secondly certain statutory procedures have been completed. The question therefore is assuming that there is no longer a non-statutory wayleave still subsisting and there is likewise no necessary wayleave in place either,

---

1   *National Grid Co* v *Craven*, see Chapter 8: section 8.4.
2   The wayleave, as a contractual licence, is subject to the same rules as any other contract. For a fuller discussion see *Chitty on Contracts* 29th ed, volume 1, Chapter 13.
3   See Chapters 6 and 7 for a full discussion of circumstances in which such events occur and the legal consequences of them.
4   See Chapters 8, 9, and 20.

is the licence-holder at that point trespassing insofar as the electric line remains on the land? If there is no non-statutory wayleave there is no longer a contract. What right then does the licence-holder have to insist upon the retention of the electric line on the land or upon the exercise of any ancillary rights at all? The right to retain the electric line upon the land is implied by the temporary continuation of the wayleave pursuant to paragraph 8 of schedule 4 to the Act.[5] The same is not necessarily true of ancillary rights.

19.5 Practical issues arising from a refusal

What a licence-holder would need in the situation outlined above would be the ability to exercise the ancillary rights pending completion of the statutory processes. While every case must be considered on its own facts and merits, it would seem reasonable to suggest that the remedy primarily relevant in such a situation is that of an injunction.

The effect of the injunction would be to restrain the landowner or occupier from interfering with the exercise of the ancillary rights pending the completion of statutory processes. With reference to the factors set out in Chapter 17, the balance of convenience would clearly be a major factor in the court's consideration. Equally a special factor would be that the temporary lack of any formal legal authority to exercise the ancillary rights would, in due course, be cured. Any damage to the landowner or occupier (it would no doubt be very strongly argued) should, in principle, be adequately compensated by an award of damages. In those circumstances an extension of the rights arising from the temporary continuation of wayleave would surely be deemed to be part of the statutory process by necessary implication.

19.6 Best practice recommendation

These issues which are replicated when it comes to considering liability for payments and use of land all arise because the exact status and legal consequences of the temporary continuation of wayleaves by statute have not been fully spelt out in the Act. We consider this further in the final chapter among our suggestions for reform. However a pragmatic solution for the present needs to be found.

---

5   See Chapter 8, section 8.6.

Our suggestion, therefore, would be that best practice indicates a negotiation of a new payment in connection with the temporary continuation of the wayleave under statute, pending completion of the necessary statutory processes such payment to be at a level consistent with an award of mesne profits had a finding in trespass been made by the court. That would underline the temporary nature of the arrangement but would also enable the landowner/occupier to recover periodic payments based on the current value of the land.

Compliance with Notice to Remove

20.1 Outline of chapter

This chapter is concerned only with circumstances which can arise in connection with an existing electric line.[1] The hypothetical legal and factual context in which we have to consider the issues raised by this chapter is this: a wayleave has come to an end by one or other of the ways previously discussed; a valid notice to remove was duly served by the grantor upon the licence-holder. This chapter is concerned with the question: when must the licence-holder comply with the notice to remove?

20.2 Suspension of notice to remove

Paragraph 8(2) of schedule 4 to the Act provides:

> the licence-holder shall not be obliged to comply with such a notice[2] except in the circumstances and to the extent provided by the following provisions of this paragraph.

Paragraphs 8(3) to 8(5) inclusive of schedule 4 to the Act prescribe the circumstances in which and the date by which an electric line must be removed in accordance with a valid notice given to the licence-holder by an owner or occupier of land. We now consider each in turn.

---

1 For equivalent circumstances relating to new lines see Chapter 21.
2 That is to say a notice to remove.

20.3 Failure to invoke statutory powers

Paragraph 8(3) contemplates the possibility that the licence-holder neither makes an application for a necessary wayleave[3] nor makes an order authorising the compulsory purchase of the land using the rights given by paragraph 1 of schedule 3 to the Act.[4] In those circumstances, the licence-holder must comply with the notice "at the end of that period".

Notice that the Act does not say "before the end of that period" but "at the end of". This suggests that the licence-holder does not become obliged to remove the electric line until the three months period from service of the notice to remove has expired. This would be consistent with the fact that an application for a necessary wayleave can be made at any time within the three month period — even on the last day.[5]

20.4 Refusal of application

Paragraph 8(4) of schedule 4 presupposes that the licence-holder has made an application for a necessary wayleave and such application has been refused by the Secretary of State. The licence-holder must remove the electric line at the end of the period of one month beginning with the date of the Secretary of State's decision or such longer period as the Secretary of State shall direct.

The obligation to remove does not, therefore, arise until the end of the period: day one of the period is the date of the decision. One month has to be read as a calendar month. Accordingly the corresponding date rule applies. Thus if the Secretary of State's decision is dated 31 March then the one month ends on 30 April. If a longer period is substituted then the same principle of calculation applies.

20.5 Refusal to confirm compulsory purchase order

Paragraph 8(5) of schedule 4 presupposes that the licence-holder makes a compulsory purchase order but such order is not confirmed

---

3   See Chapters 8–10 inclusive.
4   See Chapter 15.
5   See Chapter 9.

by the Secretary of State. The obligation to remove arises in exactly the same way as under paragraph 8(4).

20.6 Introduction of evidence relevant to an extension of the period of one month

We recommend as good practice, if more than one month is required, to introduce some evidence on this subject at the necessary wayleave hearing or the public inquiry, as the case may be, so that the Secretary of State has some information justifying a longer period.

20.7 Legal consequences of failure to remove the electric line by the due date

We now need to consider these consequences in more detail. We need to recall that the context in which we are now discussing these consequences is that the licence-holder is under a statutory obligation to remove the electric line.

What then are the legal consequences? The point has not, so far as we are aware, ever been tested by the courts. We consider this issue further in Chapter 22. Our analysis proceeds by the following stages:

- The licence-holder lawfully keeps the electric line on the land while the wayleave subsists. Failure to make a wayleave payment does not cause the wayleave to be forfeited. Forfeiture is not applicable because other procedures for termination are prescribed. The grantor in any event, has a remedy in debt.
- During the currency of a notice to remove and pending the date specified by paragraphs 8(3) to 8(5) inclusive, whichever applies, there is by definition no non-statutory wayleave in existence. It has by definition terminated. However, there is a wayleave implied by way of statutory continuation. An alternative view is that a trespass has been committed from the date the non-statutory wayleave terminated. However you view this, there should either be a continuation of wayleave payments or there should be mesne profits payable. The point is not covered by the Act. We discuss this further in Chapter 22.
- Once the due date passes and the licence-holder has not removed the electric line, then the licence-holder has committed a breach of

statutory duty by failing to remove the electric line in time. Furthermore for each day subsequently that the electric line remains in place there is an unwarranted intrusion on the land: in other words — a trespass. In such circumstances it is difficult to see why an owner or occupier of land would not in principle be entitled to seek a mandatory injunction to compel removal and compensation for any loss and damages suffered as a result of any delay in compliance. Insofar as an owner or occupier of land were seeking damages for actual loss suffered (such as diminution in the value of the land) an owner or occupier of land would be obliged to mitigate this loss by giving reasonable access, on reasonable notice, to facilitate the removal of the electric line, albeit belatedly. Subject to that qualification, however, it would be reasonable to suppose that an aggrieved landowner would be assisted by the court. Mesne profits for use of the land pending removal would also, in principle, be payable. Mesne profits would be an important remedy if non-removal of the line has not itself caused any actual loss.

A more detailed discussion of this subject is outside the scope of this book. Suffice it to say that the commentary in this paragraph serves to stress the crucial importance of having an effective grasp of when the deadline expires so far as the licence-holder's managers are concerned — and planning accordingly.

Wrongful Use of Land by Licence-holder

21.1 Outline of chapter

This chapter discusses what remedies are available to a landowner or occupier and therefore what risks would be taken by a licence-holder if an electric line were to be laid or installed on land without the owner's or occupier's consent. This issue may arise when a non-statutory wayleave expires automatically but the licence-holder has nevertheless proceeded with planned works. The issue may also arise if ancillary rights are exercised without the landowner or occupier's consent. The points made in this chapter will also be relevant if, notwithstanding the grant of a necessary wayleave, either the works associated with the laying of the electricity line or the eventual direction of the line have by mistake or for any other reason strayed outside the boundaries of the land over which the necessary wayleave has been granted.

21.2 Mesne profits or periodical payments?

As discussed in Chapter 17, mesne profits are paid for a period of trespass which is brought to an end. However, should you in fact be obtaining a wayleave payment on the basis that any use is likely to be recurrent or permanent or will possibly last for several years?

In a series of cases known collectively as "the wayleave cases"[1] the principle of "compensation for wayleave" was laid down. The question with which the courts were grappling was what is the correct principle to apply in seeking to compensate a land owner or occupier for the intermittent or frequent use of land or rights without permission. These cases did not necessarily involve a wayleave as we understand it today nor did these cases try to define a wayleave. In approaching the question, the courts developed their answer by asking what would have been payable had a proper permission been negotiated beforehand and the model of a proper permission that the Courts could take as an example was that of a wayleave. The concept of a wayleave was taken as read. Hence in *Whitwham* which was a case about tipping on land without permission of the owner, "the wayleave cases" were applied in order to assess the compensation for wrongful use by the tipper in the form of periodical payments.

In the 19th century the concept of a wayleave was also used as the example to follow when the Courts were seeking to develop principles for dealing with disputes over coal-mining activities. Referring to the so-called "wayleave cases", the Courts held that a person who passes through another person's coal mines even with his own coal must nevertheless pay compensation "as for a wayleave."[2] The same rule applied even where you are "an innocent trespasser" because you have held over after the expiration of a lease[3] but nevertheless ought to pay for having passed through another's coal mine with your own coal. If you take another person's coal wrongfully you must pay compensation for the value of the coal as severed.[4]

Both the wayleaves cases and these principles were summarised in the case of *Livingstone* v *The Rawyards Coal Company* 5 AC [1879–80] 25 where the value of the coal as severed had to be the value of the coal to an owner or occupier of land. You cannot take any of the profit made by the mining company.

Working backwards from the summary of the relevant principles in the *Livingstone* case, a wayleave fee is a fee which compensates an

---

1   *Martin* v *Porter* 5 M&W 351; *Jegon* v *Vivian* LR6 Ch App 742; *Phillips* v *Homfray* LR6 Ch App 770 as fully explained in *Whitwham* v *Westminster Brymbo Coal and Coke Company* [1896] 1 Ch Div 894 affirmed by the Court of Appeal at [1896] 2 Ch Div 538.
2   *Martin* v *Porter*, supra.
3   *Jegon* v *Vivian*, supra.
4   *Phillips* v *Homfray*, supra.

owner or occupier of land for the income which he cannot make out of the land because it is being used by the electric line and is expressed in the form of a periodic (usually annual) payment, having regard to the actual use to which the land could otherwise be lawfully put. In other words, compensation for the wrongful use of land by a licence-holder will be proportionate to the actual use made and the impact of that use upon the land in question.

21.3 Conclusion

Our suggestion is that if there is a truly temporary and intermittent use of land without consent, then mesne profits calculated on a daily basis are appropriate but if the wrongful use is permanent or intermittent but frequent then a wayleave payment is the better solution together with any compensation for the diminution in the capital value of the land wrongfully taken. The potential for clarification of this issue is something we consider in Chapter 22.

Conclusion

Utility Wayleaves — The Need for Reform

22.1 Introduction

Our proposals in this chapter build upon earlier work by others. The earliest argument for reforming the law on utilities which we have been able to find is in an RICS research paper of December 2000 produced by Norman E Hutchinson and Professor Jeremy Rowan-Robinson of Aberdeen University. Earlier work was undertaken by Barry Denyer-Green in 1999.[1] This chapter provides an overview of the historical background, and present day application of utility rights. It goes on to provide a brief summary of the various utilities outlining the legal anomalies that occur as a result of the different legislation under which they operate and the changes that have occurred with utilities generally. It concludes by arguing for reform of the legal processes if there is to be fairness and uniformity.

22.2 Historical background

Wayleave is a term welded to utility rights. It applies to the rights required for utility infrastructure. Its origins however pre-date the provision of utilities which are essential for modern living. The earliest reference to wayleaves in a law report was recorded in 1840.[2] This

---

1 B Denyer-Green, "Specific Purposes, Specific Powers: The Powers of Privatised Utilities" in Proceedings of the National Symposium on Compulsory Purchase: An Appropriate Power for the 21st Century? DETR, 1999.
2 *Dand* v *Kingscote* 6 M & W 174 or [1835–42] All ER 682.

concerned a deed entered into over 200 years previously relating to the rights to transport coal across another landowners' property. An additional term was also christened at that time, "stayleave" but this has not lasted the test of time. The essential elements in that case of determining the meaning of the wayleave was to provide sufficient rights to pass over another persons property along a haulage route of an existing railway for the purposes of carrying coal between two coal mines. The complexities of the case are not of concern as the principle was established that while the right to pass over another's land was accepted the issue to be resolved was whether this privilege was exceeded. Custom and practice allows mutual benefits to be enjoyed until one party exceeds the use contemplated by the other and when that occurs there is a wrong done for which the proper remedy is the payment of damages.

22.3 Present day application

The rights required and utilised by the major utilities; namely gas, water, electricity, sewerage and telecommunications follow on from this principle of having a right to cross over one landholding for the benefit of another. Originally, acquiring a wayleave would have been perceived as being for the public good. The taking of land to provide new transport infrastructure in the 19th century for roads, railways and canals required the original landowner's use of a specific area to be extinguished. This policy was enshrined in the Land Clauses Consolidation Act 1845 and subsequent legislation. Utilities in contrast require only limited rights allowing the landowner continuing use of some or all of the land affected by the wayleave.

The extent of rights for utilities and the associated levels of compensation for their presence have largely been unchallenged. The limited number of challenges that have taken place have been in the post privatisation era where the perceived public benefit has been upheld despite the increased commercial exploitation of utility apparatus. By way of example, telecommunication apparatus is now commonly attached to electricity transmission lines with limited change in the physical appearance of these lines. This in itself is not a novel approach as the electricity industry requires telecommunication cables to be installed between their substations in order to operate equipment remotely. In these cases the fibre optic cables are considered electricity lines but where they are for third party use additional rights are required. In this example the principle of a sufficient wayleave would be exceeded and financial recompense would be required.

The approach adopted in 1840 was to assess the extent of damages for the excessive use of the railway. In the modern equivalent example to have 12 or 72 fibres attached to or within an earth wire over 100 ft in the air cannot be conceived, in practical terms, as creating damage although additional benefit is clearly conferred on the utility company/asset owner. A separate right is required from the affected landowner and occupier but the question of financial recompense has largely been unexplored.

22.4 Sewerage

The right to construct a public sewer across private land and acquire a freehold interest can be obtained by the authority serving notice with no right of objection by the landowner concerned. This is a unique position, recognised by the Secretary of State, Michael Howard, in 1989 when promoting the Water Industry Act 1991 who said that:

> I know that in retaining the existing powers of water authorities to lay pipes on notice, we would be preserving the unique position of the water industry as the only public utility with such powers. The water industry can, however, properly be regarded in a different context from other utilities. Satisfactory water supply and sewage arrangements are essential to public health.[3]

In reality the only way to operate the functions of water companies is with electricity, for example, to power pumping stations. These are clearly essential but in the majority of cases the electricity lines used to supply these facilities are held on temporary rights. Having powerful rights to provide sanitation facilities is therefore inconsistent with the ability to deliver the solution.

22.5 Gas

The practice of gas companies is to obtain permanent rights through easements paying a landowner a percentage of the freehold value of the affected land. Additionally a lift and shift clause can be incorporated

---

3   *Hansard*, Session 1988/89, 4 July 1989, col. 180.

allowing the landowner, at some future date, to require the gas pipe to be relocated to another part of his property if it impeded development.

22.6 Telecommunications

The Telecommunications Act 1984 was the first legislation covering privatised utilities. The need to install telecommunication equipment requires the landowner's consent. Where this is not voluntarily granted a telecommunications company can seek an order of the court to install the necessary equipment. In contrast with other utilities the determining authority, the court, not only considers whether this particular apparatus should be placed on the land but additionally makes an award for compensation. This award represents the loss to the owner and also includes a consideration[4] to reflect the benefit to the telecommunication operator of utilising a particular route.

22.7 Electricity

While compulsory acquisition powers are available to electricity licence-holders invariably they place their apparatus on private land when granted a wayleave by the landowner and occupier. The Electricity Act 1989 requires that notice is given of the licence-holder's proposal, the emphasis being on the occupier primarily. Where not agreed the right to grant statutory permission in the form of a necessary wayleave is afforded to the Secretary of State for Trade and Industry. As it is not intended that a permanent right will be granted the necessary wayleave has to be for a specified term. Current practice provides for a term of 15 years with compensation being assessed on the basis of loss to the owner or occupier over that period, payable as a periodic or lump sum payment.

A significant anomaly arises where electricity lines, in place before 17 October 1972 by virtue of a terminable wayleave, and a permanent right is to be granted. The loss to the owner is then assessed only in respect of that part of the line across the landowner's property; the remaining extent of line (the works) cannot be taken into account. This is even more anomalous if adjacent pylons of the overhead line, outside

---

4   *Mercury Communications Ltd* v *London & India Dock Investments Ltd* (1995) 69 P&CR 135.

the claimant's land, reduce the value of the property in aesthetic terms and the compensation for the line across the claimant's land is based on that reduced value. It is suggested that in introducing Section 44 of the Land Compensation Act 1973 to overrule the iniquitous decision in *Edwards v Ministry of Transport*[5] created an even more unfair assessment of loss. Public works, such as highway schemes, have the CPO confirmed and notice to treat served before any work is carried out. With electricity lines the vast majority are already in place by virtue of terminable wayleaves.

An amendment could be made to both Schedule 3 and 4 to the Electricity Act 1989 providing for the assessment of compensation to take place at the date new rights are granted as if the line were to be installed. This overcomes the imbalance of situations where a landowner, having granted temporary rights for permanent structures, does not receive full compensation to reflect the loss; a situation perceived as particularly unfair where that landowner did not grant the original temporary rights.

The compensation paid to landowners for electricity lines relates to the support on the land whereas the actual use is in the wires themselves which can significantly restrict use of the land, for which a nominal payment is made. The lack of registration of these rights results in many landowners being unaware of the restrictions placed on their legal entitlement to the property. The obligation to identify the location of the utility apparatus should be placed firmly with the licence-holder. This would overcome the potential situation in which a landowner, having made reasonable enquiries, excavates within his land for development, damages a high voltage cable, for which neither rights were registered on title, nor documentary evidence provided by a previous landowner, only to receive no compensation for the temporary use of the land and, to add final insult to injury, then having to pay for the cable repair.

The area of public rights is being further extended with new entrants to the electricity distribution market being granted licences by DTI. Four such licences have been granted in addition to the licences already held by the former electricity boards at privatisation.

---

5   *Edwards v Ministry of State for Transport* [1964] 2 QB 134; [1964] 2 WLR 515.

22.8 The changing face of utility providers

Public perception takes many years to change and it is not uncommon for members of the public, and even practitioners, still to refer to the electricity board, water board and gas board despite these terms being more than 15 years expired. All utilities have characteristics peculiar to their particular requirements but the multiplicity of utilities and their multi-faceted ownership makes the need for industry specific rights less critical.

Moreover, the lack of permanent rights has not diminished the needs of utilities as temporary rights enable them to carry out their operations efficiently. In the case of electricity it is estimated that 90% of overhead lines are held on wayleaves capable of being terminated and yet some of these lines will have remained in place for over a century.

In practical terms a pipe in the ground could convey gas, water, sewerage, telecommunications, or electricity cables, entirely unseen to the naked eye. The owners of the pipe may have an interest in more than one of these purposes and they would wish to see the use maximised for commercial benefit where that is possible. The loss to the landowner is in the inability to operate over or around any pipeline but the ability to receive recompense depends significantly on the use of the pipeline for which the landowner has no secure knowledge. The example outlined of electricity lines for telecommunication purposes is quite likely to arise. In the case of fibre optics being enclosed within an earthwire, it would be impossible for a landowner to identify this separate usage. A specific power for specific purposes is therefore capable of being extended for multi utilities as the relevant legislation is inflexible to the technological changes.

22.9 Towards a utilities procedures and compensation act

The treatment of landowners when dispossessed of the entire use of their land requires uniformity. The anomalies created and accepted over time can properly be readdressed. Where permanent rights are required for utility apparatus, with sporadic rights for access and maintenance, these should be clearly set out and registered against the title to the property. Where temporary rights are required then the term should be specifically established in a statutory wayleave incorporating the level of compensation and consideration with the obligation on the utility to reapply upon expiry. Where even more transient rights are required a standard payment should be made reflecting the temporary inability to use the land for utility purposes with a de minimis wayleave fee being set. The registration of all statutory rights is capable of being put into effect.

In the 21st century emphasis is placed on freedom of information and the fair sharing of benefit and burden. A single code governing the operations of utility companies should be established in a Utilities Procedure and Compensation Act. This would provide for appropriate notice, fair hearing and recompense; irrespective of the technical specifics. The appropriate body to consider these issues is the Lands Tribunal, rather than courts or the Secretary of State. Specific issues can be dealt with by experts giving evidence whether they be engineering, valuation or planning matters.

Combining the rights of all utilities into a single codified form may appear to be difficult to achieve as technical needs will inevitably vary. However all utilities are already obliged to comply with the New Roads and Street Works Act 1991 when operating within highways. To move towards a Utilities Procedures and Compensation Act for work in private land could therefore be seen to be both achievable and long overdue.

22.10 What else might a new Act include?

We think that there are a number of improvements to the current statutory regime which could conveniently be made if such an Act were passed and which ought to be made even if the radical reform proposed above is never enacted. For convenience we take them in the order in which the issues would arise during the life of a wayleave.

22.10.1 Recording wayleaves between the parties

We think that all wayleaves should be reduced to writing in as nearly a common prescribed form as possible. Moreover we suggest that we should simplify the way we refer to wayleaves so that in substitution for the various forms of voluntary or implied wayleaves we have "non-statutory" wayleaves and in substitution for "necessary wayleaves" we have "statutory wayleaves". This is connected to the next proposal.

22.10.2 Register of wayleaves

We have already drawn attention to the fact that wayleaves are not registrable interests. Further, wayleaves may arise by implication without any formal written record of that fact. This can cause unnecessary problems for licence-holders and landowner/occupiers alike in identifying relevant interests. The problems will be most acute on transfers of land. This situation is not only inconvenient it is anomalous. It makes no sense to provide, as the Act does, that a necessary wayleave binds successors in title and then go on to provide that it shall nevertheless not be registered.

The answer is to have a register of wayleaves on which both licence-holders and landowners/occupiers must record their interests and any assignment of those interests and give at all times an address for service of documents within England and Wales. The register must state whether the wayleave is necessary or non-statutory[6] and must also state the length of the term of the wayleave, describe the land to which it relates and any provisions relating to termination (if non-statutory). The register would also name the person from whom further information and a copy of the wayleave (if express and in writing)[7] could be sought. This would permit changes in provisions for payment and ancillary rights to meet changing circumstances without requiring a change in the register. Only if one of the fundamental terms changed would an amendment to the register be required.

A licence-holder whose interest was not so registered could only benefit from any purported wayleave (whether statutory or non-statutory) if the licence-holder first served notice requiring a necessary wayleave to be granted and going through the necessary wayleave procedure if so required.

On the landowner's/occupier's side, *Welford* v *EDF Energy Network (LPN) Ltd*[8] has shown that more than one party may have an interest in land sufficient to found a claim for statutory compensation following the grant of a necessary wayleave. Other examples would include that of a landlord and tenant each having a distinct legal interest in the same land. An advantage of a wayleaves register is that

---

6 If the first proposal were accepted the choice would in fact be between "statutory" or "non-statutory".
7 If the first proposal were accepted the only form of non-statutory wayleave would be express and written.
8 [2007] EWCA Civ 293.

all such interests could be noted just as multiple interests can be noted on an insurance policy.

It would be a necessary precondition to being entitled to be served with an application for a necessary wayleave and to being entitled to compensation arising from the grant of a necessary wayleave or disturbance to moveables that a landowner and/occupier should first have registered his legal interest in the land which is subject to the wayleave. Whether in any particular case a person so interested has suffered any loss which falls to be compensated in accordance with the compensation code would be a matter to be determined by the Lands Tribunal.

22.10.3 Compensation for wrongful use of land during the continuance of a wayleave

The term wrongful use while venerable in legal circles carries with it a sense of personal misconduct. As discussed in earlier chapters, wrongful use can mean a number of things: stepping outside the boundaries of the wayleave; intensification of ancillary rights; continuing to exercise wayleave rights even though the wayleave has actually expired. In our view if a licence-holder uses land wrongfully in any of these ways then it should be clear that the proper remedy is first payment of an additional wayleave fee proportionate to the nature and the length of the use and if any actual capital loss is suffered, a payment assessed in accordance with the compensation code. In the event of dispute the matter would be referred to the Lands Tribunal to be determined either in accordance with the written submissions procedure, if there were no dispute of fact only valuation or in accordance with the simplified procedure if there is an issue of fact to be determined.

22.10.4 Explicit sanctions for failure to remove an electric line

At present there is nothing in the Act or in case-law which defines by what long-stop date a licence-holder must remove an electric line if a notice to remove is served and no application is made for a necessary wayleave. Furthermore, there is no explicit authority stating what remedy is available to a landowner or occupier aggrieved by a failure to remove. In our view these are glaring omissions in the law which ought to be corrected. The ordinary common law remedies are sufficient for the purpose. There would however need to be specific provision for the terms of any order to remove having regard to the practical circumstances. In financial terms, mesne profits equivalent to the last agreed wayleave payment should be payable pending removal and if any capital loss can also be shown then subject to the compensation code that can be assessed by the Lands Tribunal.

22.10.5 Payments and ancillary rights during a statutory continuation

The law is far from clear on this point. In our view analogously with the Landlord and Tenant Act 1954, if a licence-holder applies for a necessary wayleave, then all the terms and conditions in favour of each party should apply during the statutory continuation without prejudice to that application.

22.10.6 Applications for a necessary wayleave

In a society where our dealings between each other and with the state are protected by human rights legislation and where the proper conduct of our legal system is overseen by a Ministry of Justice, it is wholly unacceptable that such applications are remitted to a minister from whose decision there is no right of appeal. Furthermore it is inconsistent with the reforms to civil procedure and the Lands Tribunal Rules that such applications are governed by a procedural code which does not mandate a reciprocal exchange of documents and evidence.

In our view applications for a necessary wayleave should be remitted to the Lands Tribunal. This would have three advantages. There would be an appeal from any decision on a point of law. The question of compensation could be determined simultaneously with the substantive decision if a wayleave were granted. (This has the additional merit of enabling the Lands Tribunal to look at the question of expediency in the context of losses likely to be suffered by the landowner or occupier in comparison to the prejudice to the licence-holder arising from its inability to fulfil its statutory duties. It would also substantially reduce delay: not only would there be one process rather than potentially two; there would be a proper code for the management of the dispute which could be enforced.) Finally, the Lands Tribunal would have the power to award costs to either party if in its discretion it considered that appropriate. From the Treasury's point of view there would be this benefit: the remission of this further jurisdiction to the Lands Tribunal would enable fees to be charged.

There would be no loss of expertise available for the determination of such applications insofar as the DTI inspectors could sit as assessors in the Lands Tribunal.

Contrary, in our opinion, to the judicial observations in *British Waterways Board* v *London Power Networks plc*,[9] the DTI takes the view that the Secretary of State can only grant the actual wayleave for which the licence-holder applies. In our view, the landowner or occupier should have the right to make a counter-application for a wayleave on alternative terms. If such applications were remitted to the Lands Tribunal then the ability to bring a counterclaim could be foreshadowed by open offers prior to proceedings. The same cost consequences as section 4 of the 1961 Act applies to pre-application offers of compensation could be made to apply equally to pre-application offers of terms for a wayleave.

We have drawn attention to a problem which arises if a licence-holder is obliged to withdraw an application for a wayleave to facilitate settlement. If such applications were remitted to the Lands Tribunal then there would be ample discretion to grant stays of proceedings on terms to permit negotiations. In any event in our view if the legislative framework remains unchanged then the Secretary of State should direct the DTI to permit one stay of a wayleave hearing provided it does not exceed three months.

Chapter 16 draws attention to the various sources for the rules under which the Lands Tribunal operates. In our view it is both necessary and expedient that they be found only in an updated set of Lands Tribunal Rules and not in the 1949 Act supplemented only by such additional points as are contained in the Practice Directions. At present there is much duplication between the two because rather than enact new legislation, the government has tried to bring the current Lands Tribunal Rules into line with the Civil Procedure Rules by means of the Practice Directions only.

22.10.7 *Welford v EDF Energy Network (LPN) Ltd*[10]

We do not dispute the actual decision of the case based on the facts as found by the Lands Tribunal. However the Court of Appeal clearly

---

9   [2003] 1 All ER 187.
10 [2007] EWCA Civ 293.

had some reservations about that finding. In our view if the test of sufficient connection with the land in the case of loss of profits has to be determined by whether a business has been commenced on the land, then the facts to be proved should be no different from those necessary to prove that a person is in occupation of land for the purposes of a business under the provisions of the Landlord and Tenant Act 1954.

Appendix 1

Map of Area Electricity Boards 1947–1990

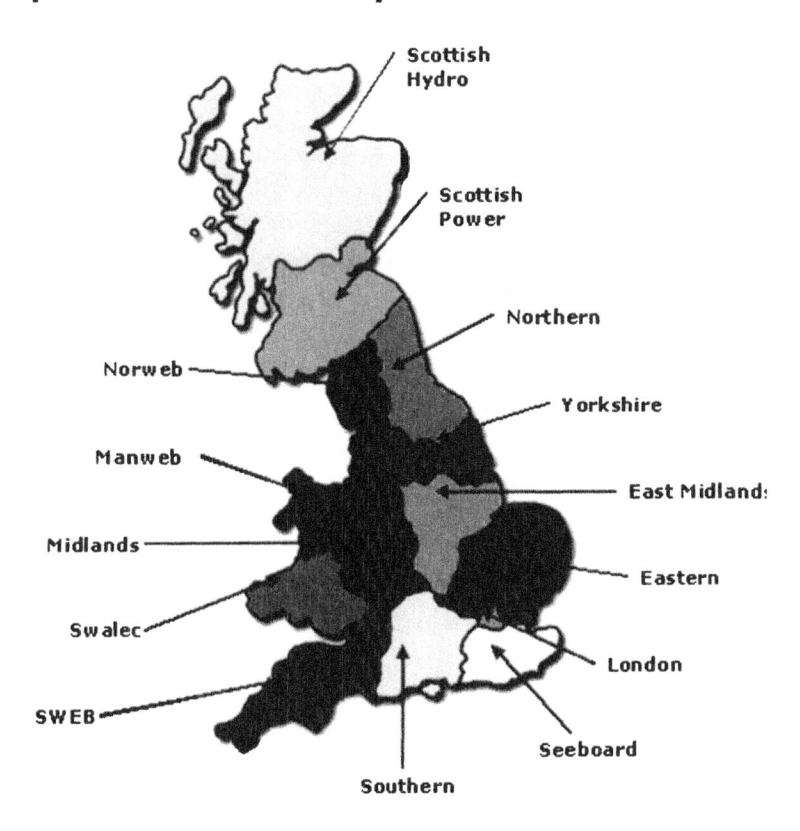

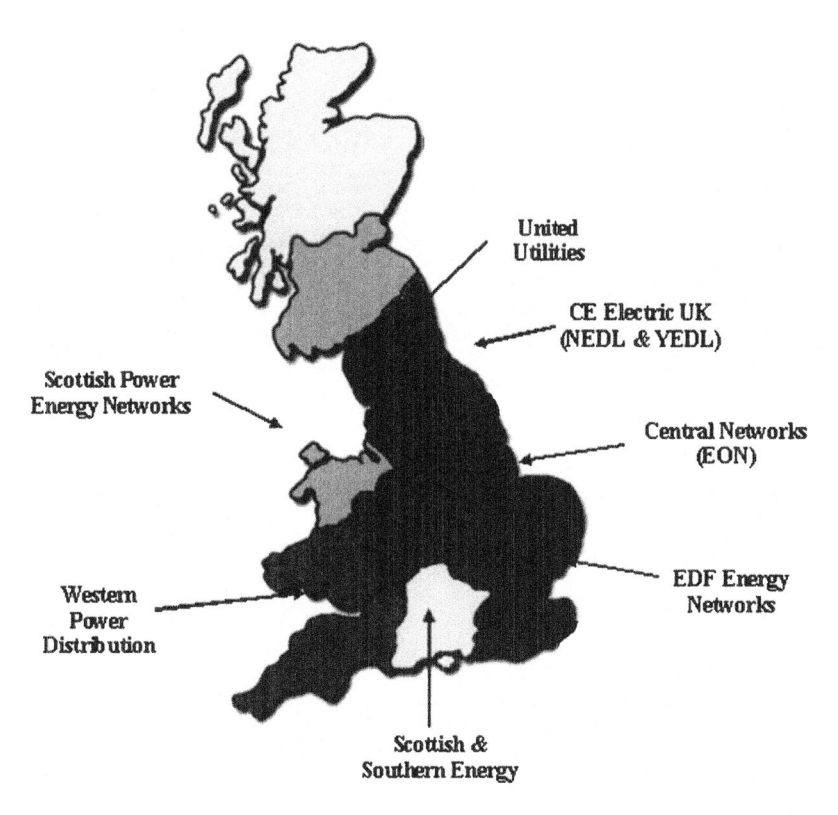

Appendix 2

Map of Distribution Network Operators 2006

Appendix 3

Table of Area Electricity Boards, Regional Companies and Distribution Network Operators

Area Electricity Board 1947–1990	Regional Electricity Company 1990	Distribution Network Operator 2007
Eastern Electricity Board	Eastern Electricity plc	EDF Energy Networks (EPN) plc
East Midlands Electricity Board (EMEB)	East Midlands Electricity plc	Central Networks East plc (E.ON)
London Electricity Board (LEB)	London Electricity plc	EDF Energy Networks (LPN) plc
Merseyside and North Wales Electricity Board (MANWEB)	Manweb plc	SP Manweb plc (Scottish Power Energy Networks)
Midlands Electricity Board (MEB)	Midlands Electricity plc	Central Networks West plc (E.ON)
North Eastern Electricity Board (NEEB)	Northern Electricity plc	Northern Electric Distribution Limited (CE Electric UK)
North Western Electricity Board (NORWEB)	Norweb plc	United Utilities Electricity plc
Southern Electricity Board (SEB)	Southern Electric plc	Southern Electric Power Distribution plc (Scottish and Southern Energy plc)
South Eastern Electricity Board (SEEBOARD)	Seeboard plc	EDF Energy Networks (SPN) plc
South Wales Electricity Board	South Wales Electricity plc	Western Power Distribution (South Wales) plc
South Western Electricity Board (SWEB)	South Western Electricity plc	Western Power Distribution (South West) plc
Yorkshire Electricity Board (YEB)	Yorkshire Electricity plc	Yorkshire Electricity Distribution plc (CE Electric)

Appendix 4

Illustration of the Generation, Transmission and Distribution of Electricity

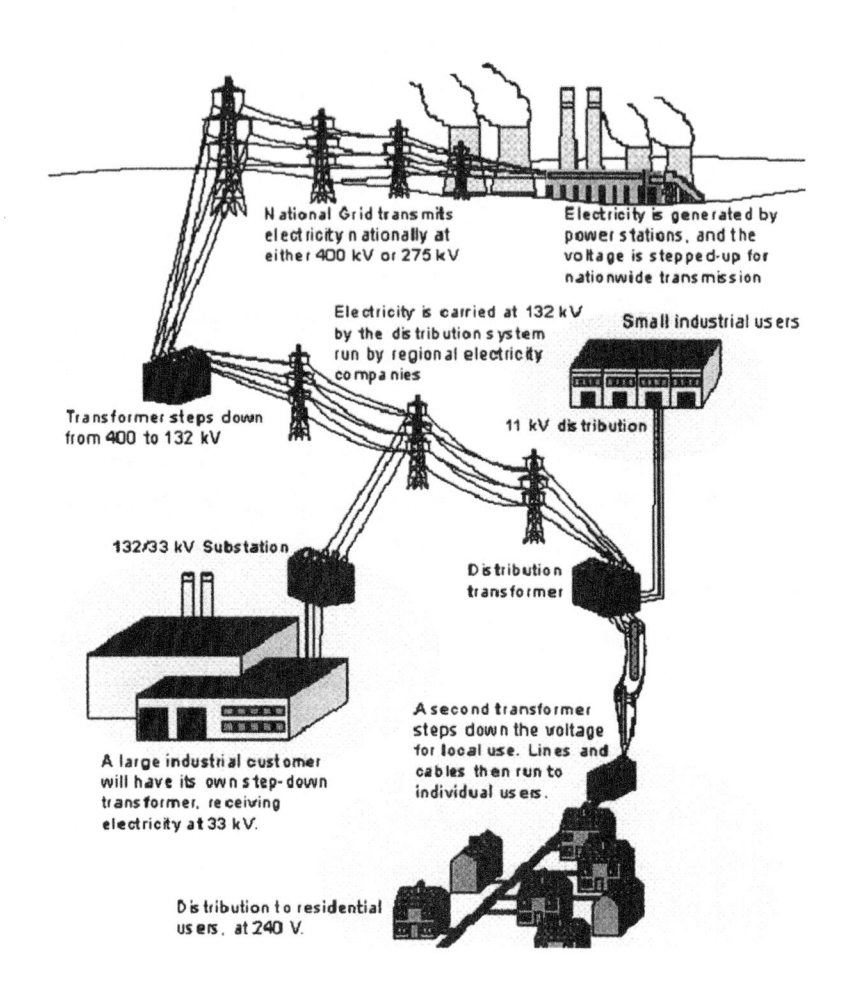

Electricity Wayleaves, Easements and Consents

National Grid transmits electricity nationally at either 400 kV or 275 kV

Electricity is generated by power stations, and the voltage is stepped-up for nationwide transmission

Electricity is carried at 132 kV by the distribution system run by regional electricity companies

Small industrial users

Transformer steps down from 400 to 132 kV

11 kV distribution

132/33 kV Substation

Distribution transformer

A large industrial customer will have its own step-down transformer, receiving electricity at 33 kV.

A second transformer steps down the voltage for local use. Lines and cables then run to individual users.

Distribution to residential users, at 240 V.

© National Grid 2005

Reproduced by kind permission of National Grid

Appendix 5

Typical Suspension/High Voltage Towers
Typical Suspension Towers

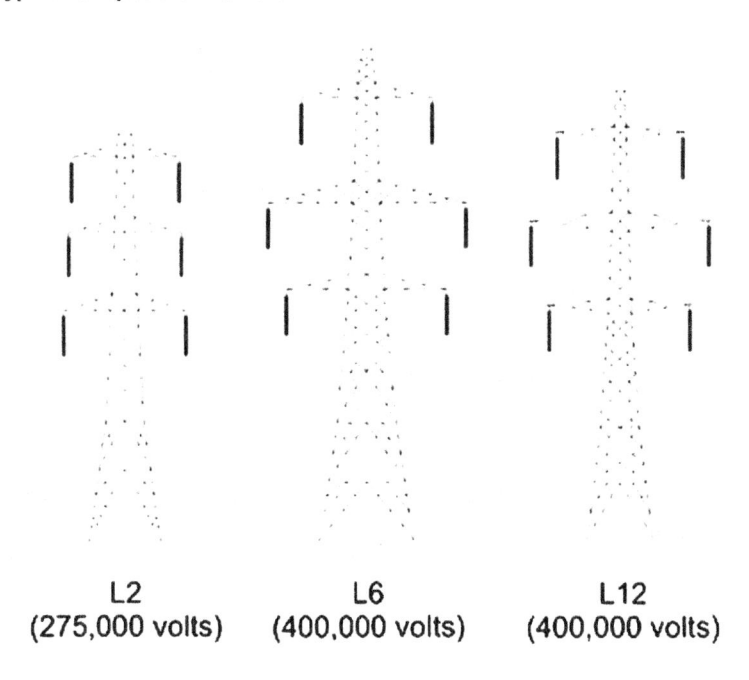

L2
(275,000 volts)

L6
(400,000 volts)

L12
(400,000 volts)

Source: National Grid — Planning and Amenity Aspects of High Voltage Transmission Lines and Substations — Information for Planning Authorities and Developers — Figure 1

© 2001 The National Grid Company plc. Reproduced by kind permission of National Grid

Electricity Wayleaves, Easements and Consents

Typical High Voltage Towers in a 400kV Route

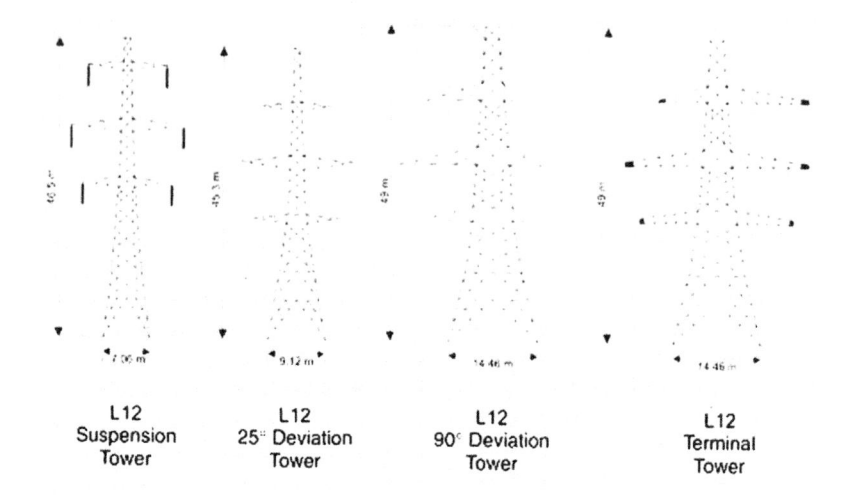

| L12 Suspension Tower | L12 25° Deviation Tower | L12 90° Deviation Tower | L12 Terminal Tower |

Source: National Grid — Planning and Amenity Aspects of High Voltage Transmission Lines and Substations — Information for Planning Authorities and Developers — Figure 2

© 2001 The National Grid Company plc. Reproduced by kind permission of National Grid

Appendix 6

Typical Double Circuit Overhead Line

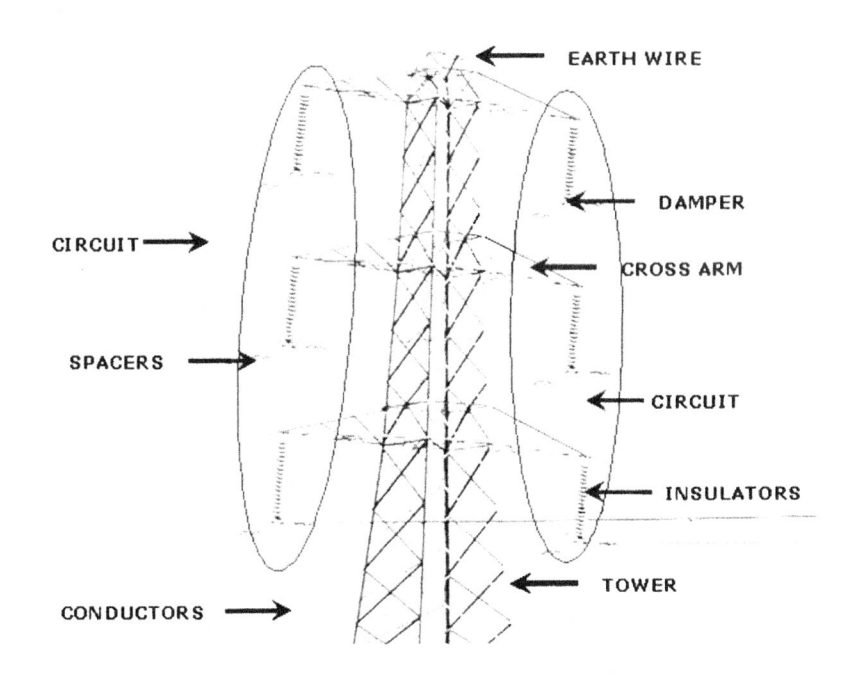

← EARTH WIRE
← DAMPER
CIRCUIT →
← CROSS ARM
SPACERS →
← CIRCUIT
← INSULATORS
← TOWER
CONDUCTORS →

Source: Hamer Associates Ltd

Appendix 7

Typical Single Circuit Overhead Line

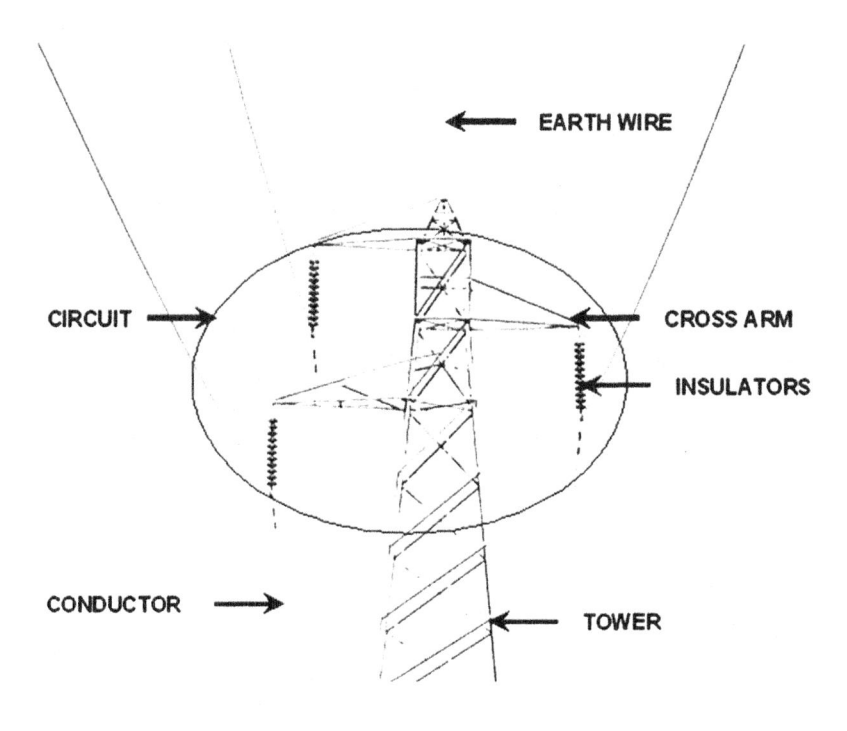

Source: Hamer Associates Ltd

Appendix 8

Glossary of Terms Relating to Electricity Apparatus[1]

Angle tower   Towers carrying one or two circuits where the route of the line changes direction. These towers are bulkier than suspension towers and require more extensive foundations due to the stresses imposed by the tension of the conductors. Angle towers for 275,000 and 400,000 volts overhead lines would be particularly bulky and in all cases rarely with an angle greater than 90° capable of being constructed. Most towers have a tolerance of deviation with the vast majority allowing for 10° or 30° deviation in the angle of the line conductors.

Application factor   The distance (dependent upon working situation) which must be added to safety distance to determine working and access clearance.

Basic electrical clearances   Clearance to earth ascribed to air insulation for each system voltage. Basic electrical clearances do not include any additions for constructional tolerances, wind effects, etc.

---

1   Source: ESI Standard 43-8 and Hamer Associates Ltd.

Cable	A conductor, or assembly of conductors, which are effectively insulated.
Catenary	Suspended line creating a curve formed by uniform wire hanging freely from two points not in a vertical line.
Conductor	The wire strung between pylons, used for transmitting electricity. The conductor is the part of the cable which carries the electricity and is almost exclusively made of copper or aluminium.
Controlled zone	Inside an enclosure efficiently protected by fencing not less than 2.4m (8ft) in height or other means so as to prevent access to the electric lines and apparatus therein by any unauthorised person.
Creep	Non elastic stretch of a conductor. This consists of two parts — bedding down of the strands and the long term stretch of conductor material.
Crossarm	The horizontal lattice steelwork of pylons holding the insulators.
Damper	Metal device fixed to conductors to avoid insulator damage in windy conditions.
Diamond crossing	Where an overhead line has to cross the route of another overhead line a typical arrangement is to run the conductors, where there are two circuits, each side of a tower of the existing line. To achieve this the new line has to go beneath the existing line on smaller towers or gantries and the new conductors would go each side of the existing tower and then connect from smaller towers or gantries. The continuation of the new line producing, in plan view, a diamond shape.
Distribution	Defined as to distribute by means of a distribution system which consists (wholly or mainly) of low voltage lines and electrical plant

Glossary of Terms

	and is used for conveying electricity to any premises or to any other distribution system.
Effectively insulated conductor	A line conductor which is insulated for the working voltage and, where necessary, sheathed to afford mechanical protection.
Gabion walls	Cylinder of wicker or woven metal bands to be filled with earth or stones for use in engineering or fortification — hence line of gabions.
High voltage	An electric line of a nominal voltage exceeding 132,000 volts.
Insulator	Used to attach the conductors to the pylons preventing electrical discharge to the steelwork. Usually made from porcelain, glass or ceramic units joined together to form an insulator string.
Jumper connection	A connection at a support from a phase conductor to another conductor or to a terminal on transformers, switchgear, fuse gear, line taps, etc at support.
Lightly insulated conductor	A line conductor which is insulated against momentary phase to phase or phase to earth contact and must be considered as a bare conductor for ground clearance purposes.
Line conductor	A conductor used, or to be used, for conveying a supply of electricity. A line conductor is deemed to include a through jumper.
Normal use of land	The type of work or activity which is likely to occur on or over a particular piece of land or water.
Overhead line apparatus	In the open air and above ground level coming within the ambit of the overhead line regulations.
Safety distance	Distance from nearest exposed conductor or from an insulator supporting a conductor which must be maintained to avoid danger.

Sealing end platform These platforms are installed at terminal tower positions where the cables are connected to the overhead line conductors. These platforms are generally installed at a height of at least seven metres above ground level and support the connecting equipment from the insulators of the overhead line conductors to the underground cables. Where the line is double circuit then a platform would be installed each side of the tower and typically for a single circuit line would be installed on one side only. In urban areas or areas susceptible to vandalism a stone mesh guard would be installed around the equipment on the platform to minimise damage to the insulators and equipment.

Section tower Towers carrying one or two circuits where the continuous run of conductors stops at the tower requiring the tension each side to be balanced with a loop of conductors providing electricity continuity. The sections of conductor are often changed at angle tower positions or generally every three towers.

Spacer A fixing attached between conductors to stop them from touching each other.

Specified maximum temperature The likely maximum temperature of the conductor resulting from a combination of climatic conditions and the rated electrical load under operating conditions.

Suspension tower Towers carrying one or two circuits in a straight line.

System voltage The nominal root mean square phase to phase voltage of a three-phase alternating current system.

Terminal tower Where the conductors of the line terminate at a tower position either within a sub station where the conductors then connect to overhead switchgear or where the conductors connect to underground cable. These towers are generally bulkier in their structure but often of a lower height than suspension towers to minimise the connection between the underground cable at the sealing end platform or into the switchgear within a sub station compound.

Transmission system Defined as a system which consists (wholly or mainly) of high voltage lines and electrical plant and is used for conveying electricity from a generating station to a sub station, from one generating station to another or from one sub station to another

Wire A wire which is not designed to convey electricity, but which is attached to a support carrying line conductors, eg flying stay wire.

Appendix 9

Clearance to Trees

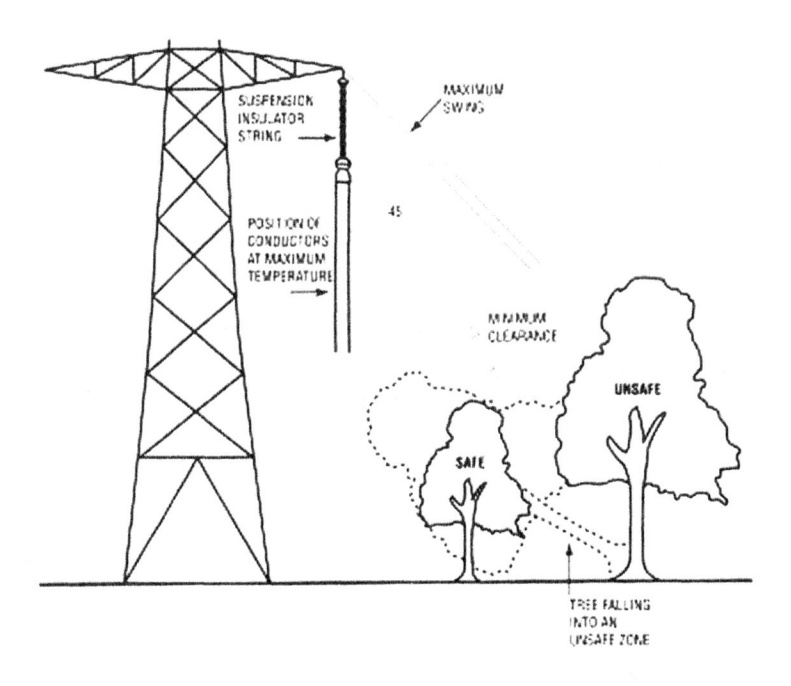

Source: National Grid — Planning and Amenity Aspects of High Voltage Transmission Lines and Substations — Information for Planning Authorities and Developers — Figure 5

© 2001 The National Grid Company plc. Reproduced by kind permission of National Grid

Appendix 10

Clearance to Objects (on which a person can stand

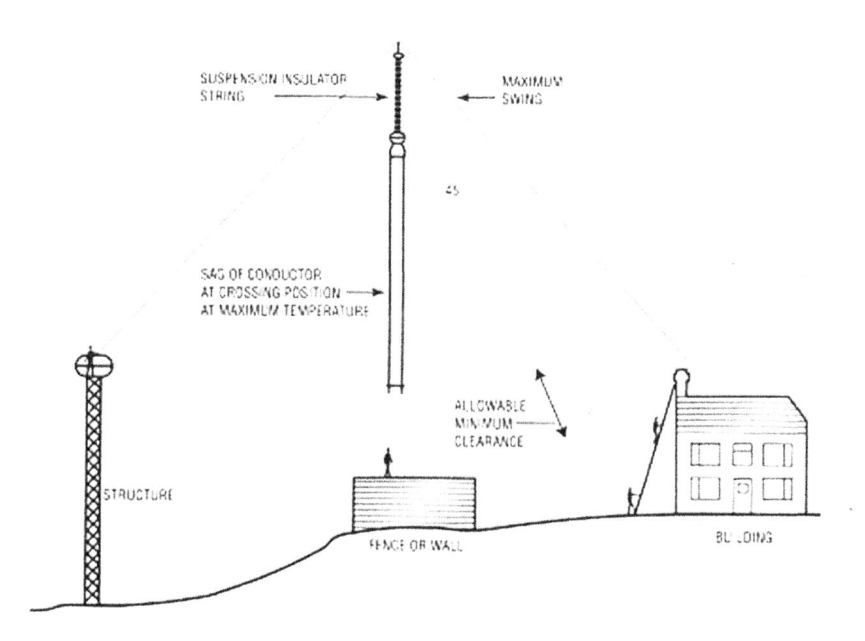

Source: National Grid — Planning and Amenity Aspects of High Voltage Transmission Lines and Substations — Information for Planning Authorities and Developers — Figure 4

© 2001 The National Grid Company plc. Reproduced by kind permission of National Grid

Appendix 11A

Flowchart of the Necessary Wayleave Process

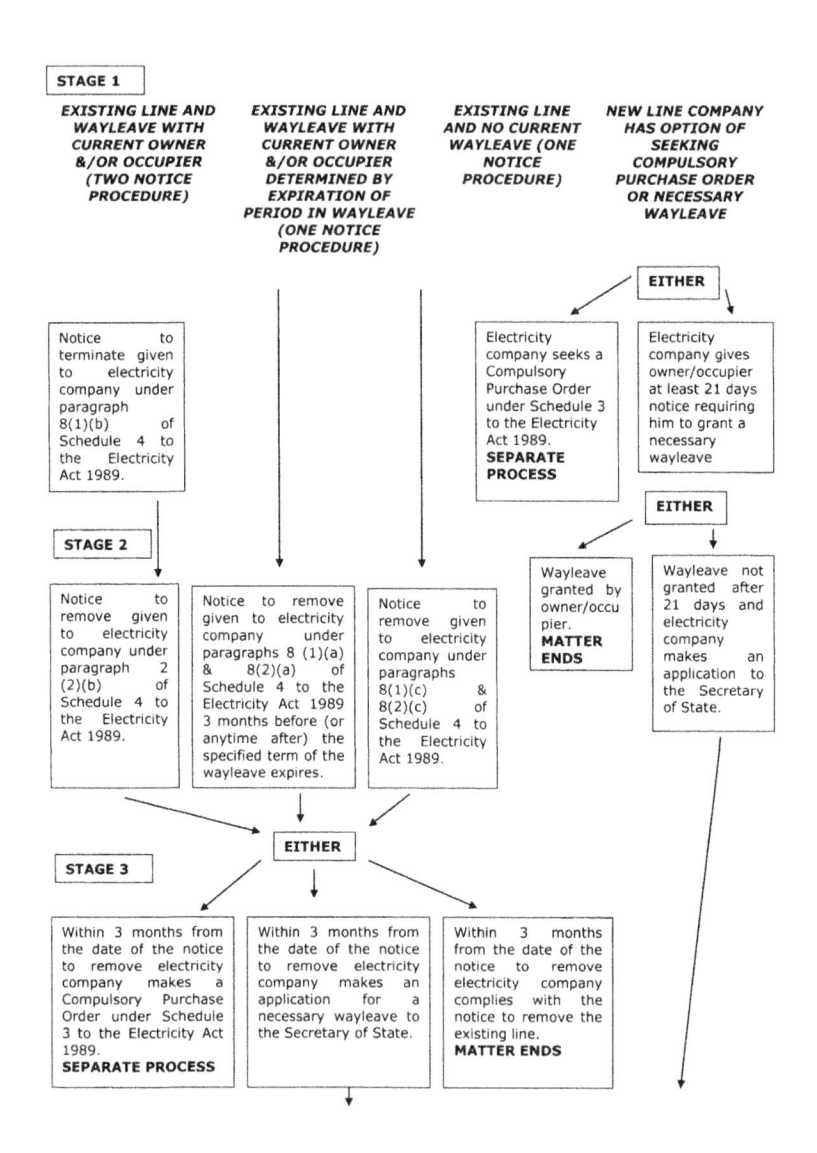

Electricity Wayleaves, Easements and Consents

STAGE 1

EXISTING LINE AND WAYLEAVE WITH CURRENT OWNER &/OR OCCUPIER (TWO NOTICE PROCEDURE)	EXISTING LINE AND WAYLEAVE WITH CURRENT OWNER &/OR OCCUPIER DETERMINED BY EXPIRATION OF PERIOD IN WAYLEAVE (ONE NOTICE PROCEDURE)	EXISTING LINE AND NO CURRENT WAYLEAVE (ONE NOTICE PROCEDURE)	NEW LINE COMPANY HAS OPTION OF SEEKING COMPULSORY PURCHASE ORDER OR NECESSARY WAYLEAVE
Notice to terminate given to electricity company under paragraph 8(1)(b) of Schedule 4 to the Electricity Act 1989.		Electricity company seeks a Compulsory Purchase Order under Schedule 3 to the Electricity Act 1989. **SEPARATE PROCESS**	**EITHER** — Electricity company gives owner/occupier at least 21 days notice requiring him to grant a necessary wayleave

STAGE 2

| Notice to remove given to electricity company under paragraph 2(2)(b) of Schedule 4 to the Electricity Act 1989. | Notice to remove given to electricity company under paragraphs 8(1)(a) & 8(2)(a) of Schedule 4 to the Electricity Act 1989 3 months before (or anytime after) the specified term of the wayleave expires. | Notice to remove given to electricity company under paragraphs 8(1)(c) & 8(2)(c) of Schedule 4 to the Electricity Act 1989. | **EITHER** — Wayleave granted by owner/occupier. **MATTER ENDS** / Wayleave not granted after 21 days and electricity company makes an application to the Secretary of State. |

EITHER

STAGE 3

| Within 3 months from the date of the notice to remove electricity company makes a Compulsory Purchase Order under Schedule 3 to the Electricity Act 1989. **SEPARATE PROCESS** | Within 3 months from the date of the notice to remove electricity company makes an application for a necessary wayleave to the Secretary of State. | Within 3 months from the date of the notice to remove electricity company complies with the notice to remove the existing line. **MATTER ENDS** |

274

Flowchart of the Necessary Wayleave Process

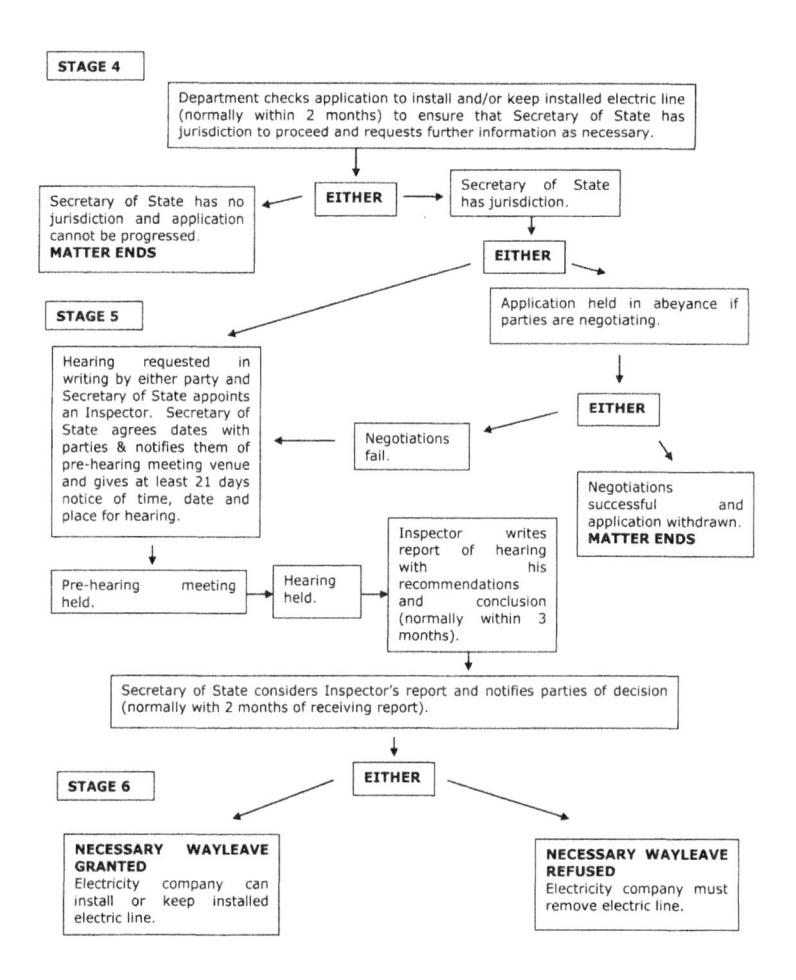

STAGE 4

Department checks application to install and/or keep installed electric line (normally within 2 months) to ensure that Secretary of State has jurisdiction to proceed and requests further information as necessary.

↓

EITHER

- Secretary of State has no jurisdiction and application cannot be progressed. **MATTER ENDS**
- Secretary of State has jurisdiction.

↓

EITHER

- Application held in abeyance if parties are negotiating.

  ↓

  EITHER
  - Negotiations fail.
  - Negotiations successful and application withdrawn. **MATTER ENDS**

STAGE 5

Hearing requested in writing by either party and Secretary of State appoints an Inspector. Secretary of State agrees dates with parties & notifies them of pre-hearing meeting venue and gives at least 21 days notice of time, date and place for hearing.

↓

Pre-hearing meeting held. → Hearing held. → Inspector writes report of hearing with his recommendations and conclusion (normally within 3 months).

↓

Secretary of State considers Inspector's report and notifies parties of decision (normally with 2 months of receiving report).

↓

EITHER

STAGE 6

- **NECESSARY WAYLEAVE GRANTED**
  Electricity company can install or keep installed electric line.
- **NECESSARY WAYLEAVE REFUSED**
  Electricity company must remove electric line.

Appendix 11B

Contents of an Application for a Necessary Wayleave: General Requirements

The following information should be included in all applications for a necessary wayleave:

- Details of the owner and/or occupier.
- The location of the land.
- Details of the electric line and apparatus in question.
- A statement as to whether the application is for one or more necessary wayleaves and the number of lines covered by each necessary wayleave applied for.
- A statement as to whether the necessary wayleave application is to install a new electric line under paragraph 6(1)(a) of schedule 4 to the Act or to keep an electric line installed under paragraph 6(2) of schedule 4 to the Act.
- A plan/map (of reasonable scale) but ideally on A4/A3 size paper to make photocopying easier, clearly detailing:

    - the owner and/or occupier's affected land boundaries
    - the electric lines and apparatus in question crossing that land (showing the distinction between lines placed or to be placed under or over ground) and
    - other electric lines crossing the land but not subject to the necessary wayleave application

- confirmation of the state of play on negotiations (ie whether they are continuing and the necessary wayleave application should be held in abeyance or whether a hearing is required).

[Note when assembling information in response to the requirements set out above, attention should also be paid to the commentary in Chapter 9, sections 9.9 to 9.11 inclusive.]

Appendix 11C

Contents of an Application for a Necessary Wayleave: Additional Information in Respect of an Existing Line

[Note: this information is to be added to the information provided in response to the requirements set out in Appendix 11B.]

- A copy of the written notice from the owner and/or occupier to remove the electric line from the land.
- A copy of any previous written notice to terminate from the owner and/or occupier (if applicable).
- Any relevant express wayleave agreements [this information is relevant to deterring which part of paragraph 8(1) of schedule 4 to the Act applies, usually a necessary wayleave application is made in accordance with either paragraph 8(1)(a) or 8(1)(b) of Schedule 4 to the Act].

If you are unable to locate any previous wayleave agreement for the electric line in question then the terms need to be established by reference to the following:

- whether or not circumstances have arisen which result in a contractual arrangement being implied by the action of parties and all documents avoiding those circumstances
- whether any record exists of wayleave payments having been made in the past for the electric line

- whether any record exists of consent being given for the installation of the line
- whether there are any records of previous objections or disputes regarding the electric line and
- how long the line has been installed.

Appendix 11D

Contents of an Application for a Necessary Wayleave: Additional Information in Respect of a New Line

[Note this information is to be added to the information provided in response to the requirements set out in Appendix 11B.]

Confirmation that:

- at least 21 days notice has been given to the owner/occupier requiring him to give the necessary wayleave in accordance with paragraph 6(1)(b) schedule 4 to the Act (a copy of the notice should be attached)
- except for electric lines to be placed underground, the land is not covered by a dwelling as defined in paragraph 6(8) of schedule 4 to the Act and
- except in the case of electric lines to be placed underground, confirmation that planning permission is not in force for a dwelling to be constructed.

Appendix 12

Form B (Type II)

> Notes: This Form is for use in connection with an application for the consent of the Secretary of State for Trade and Industry's consent under section 37 of the Electricity Act 1989 to install and keep installed above ground an electric line of a nominal voltage of 132 kilovolts or more. The Form should be sent in quadruplicate to each County (where there is one) and District/Borough Council in whose area the proposed development would be situated. Where the Form is sent to more than one District/Borough Council in a county the County Council should be sent an additional copy for each such additional District/Borough Council

DETAILS OF APPLICANT

Name: LICENCE-HOLDER

Address:

Tel:

PART I

Applicant's reference: Date:
To the Chief Executive Council

Dear Sir

Electricity Act 1989

Application is being made to the Secretary of State for Trade and Industry for his consent to development described overleaf. The Secretary of State will at the same time be requested to direct that planning permission for this development shall be deemed to be granted. The consent and the direction may be given subject to conditions.

To assist the Secretary of State to determine the application:

(i) the District Council is requested to return to me two copies of this Form with Part I Certificate and Part II completed and signed and to send one completed and signed copy to the County Council (where there is one),
AND

(ii) the County Council (where there is one) is requested to return to me two copies of this Form with Part I Certificate only completed and signed and to send one completed and signed copy to each District Council in whose area the development would be situated.

Department of the Environment Circular 14/90 (Welsh Office Circular 20/90) (Department of Energy Circular 1/90) describes this procedure and the reason for it.

Yours faithfully

For and on behalf of the applicant

CERTIFICATE

(To be completed by or on behalf of BOTH County AND District/Borough Councils)

The   County/District/Borough Council

(i) *object on the grounds set out below/have no objection to make to the development described overleaf;

Form B (Type II)

(ii) *request/ do not request that a public inquiry be held pursuant to paragraph 2 of Schedule 8 to the Electricity Act 1989 before the Secretary of State reaches his decision on the application.

Dated   Signed

*Delete as appropriate   Designation

                On behalf of the County/District/Borough Council

                [Reasons for objections]

PARTICULARS OF PROPOSED DEVELOPMENT AND REPRESENTATIONS

[To be completed by the applicant]

Application is being made

(a) for consent under section 37 of the Electricity Act 1989 to install or keep installed an electric line above ground.

(b) for a direction under section 90(2) of the Town and Country Planning Act 1990 that planning permission for the proposed development be deemed to be granted.

1. Particulars of proposed development. (These particulars should be accompanied by such plans as may be necessary to enable the local planning authority to identify the land affected by the proposals and to appreciate the nature and extent of the proposed development and by a copy of the environmental statement if the applicant has prepared one.)

2. Particulars of any representations or objections which have been made to the applicant.

3. Particulars of the applicant's compliance with his duty under paragraph 1 of Schedule 9 to the Electricity Act 1989.

Date   For and on behalf of the applicant

Note: This Part to be completed,   Signed
dated and signed before submitting   Designation
to the local planning authority.

PART II — INFORMATION AND OBSERVATIONS

(To be completed by the District/Borough Council)
                                            Planning Reference No.

1. Names of interested parties consulted as to the proposals with details of any observations received.

2. Particulars of any representations which have been made to the local planning authority objecting to the proposals.

3. Have any general or specific restrictions been imposed by any authority on development of the land affected by the proposed development?

4. Does the proposed development involve the demolition, alterations or extension of a building of special architectural or historic interest included in a list compiled or approved under section 1 of the Planning (Listed Buildings and Conservation Areas) Act 1990?

5. Does the local planning authority agree that the proposed development should be approved by the Secretary of State for Trade and Industry as described? (If the answer is no, please answer question 6.)

6. Would the local planning authority be prepared to agree that the proposed development should be approved subject to

Form B (Type II)

modifications or conditions? (If so, specify the modifications or conditions proposed and state whether they are acceptable to the applicant). (Note: The precise form of any modifications or conditions subject to which the consent or direction is given is a matter for the Secretary of State, who will however have regard to the form of words used.)

Dated   Signed

                                    Designation

On behalf of the   Council

(Local planning authority for the area in which the proposed development is to be carried out)

Two completed copies of this Form, both signed, should be returned to the applicant for submission by them to the Department of Trade and Industry.

Appendix 13

Form B (Type III)

> Notes: This Form is for use in connection with an application for the Secretary of State for Trade and Industry's consent under **section 37** of the Electricity Act 1989 to install and keep installed above ground an electric line of a nominal voltage of less than 132 kilovolts. The Form should be sent **in triplicate** to each District/Borough Council in whose area the proposed development would be situated.

DETAILS OF APPLICANT

Name: LICENCE-HOLDER

Address:

Tel:

PART I

Applicant's reference: Date:

To the Chief Executive   Council

Dear Sir

Electricity Act 1989

Application is being made to the Secretary of State for Trade and Industry for his consent to development described overleaf. The Secretary of State will at the same time be requested to direct that planning permission for this development shall be deemed to be granted. The consent and the direction may be given subject to conditions.

To assist the Secretary of State to determine the application District/Borough Council is requested to return to me two copies of this Form with Part I Certificate and Part II completed and signed.

Department of the Environment Circular 14/90 (Welsh Office Circular 20/90) (Department of Energy Circular 1/90) describes this procedure and the reason for it.

Yours faithfully

For and on behalf of the applicant

CERTIFICATE

(To be completed by or on behalf of the District/Borough Councils)

The   District/Borough Council

(i) *object on the grounds set out below/have no objection to make to the development described overleaf;
(ii) *request/ do not request that a public inquiry be held pursuant to paragraph 2 of Schedule 8 to the Electricity Act 1989 before the Secretary of State reaches his decision on the application.

Dated   Signed

*Delete as appropriate   Designation

                                 On behalf of the District/Borough Council
                                 [Reasons for objections]

Form B (Type III)

PARTICULARS OF PROPOSED DEVELOPMENT AND REPRESENTATIONS

[To be completed by the applicant]

Application is being made

(a) for consent under section 37 of the Electricity Act 1989 to install or keep installed an electric line above ground.

(b) for a direction under section 90(2) of the Town and Country Planning Act 1990 that planning permission for the proposed development be deemed to be granted.

1. Particulars of proposed development. (These particulars should be accompanied by such plans as may be necessary to enable the local planning authority to identify the land affected by the proposals and to appreciate the nature and extent of the proposed development and by a copy of the environmental statement if the applicant has prepared one.)

2. Particulars of any representations or objections which have been made to the applicant.

3. Particulars of the applicant's compliance with his duty under paragraph 1 of Schedule 9 to the Electricity Act 1989.

Date   For and on behalf of the applicant

Note: This Part to be completed,   Signed
dated and signed before submitting   Designation
to the local planning authority.

PART II — INFORMATION AND OBSERVATIONS

(To be completed by the District/Borough Council)
Planning Reference No.

1. Names of interested parties consulted as to the proposals with details of any observations received.

2. Particulars of any representations which have been made to the local planning authority objecting to the proposals.

3. Have any general or specific restrictions been imposed by any authority on development of the land affected by the proposed development?

4. Does the proposed development involve the demolition, alterations or extension of a building of special architectural or historic interest included in a list compiled or approved under section 1 of the Planning (Listed Buildings and Conservation Areas) Act 1990?

5. Does the local planning authority agree that the proposed development should be approved by the Secretary of State for Trade and Industry as described? (If the answer is no, please answer question 6.)

6. Would the local planning authority be prepared to agree that the proposed development should be approved subject to modifications or conditions? (If so, specify the modifications or conditions proposed and state whether they are acceptable to the applicant). (Note: The precise form of any modifications or conditions subject to which the consent or direction is given is a matter for the Secretary of State, who will however have regard to the form of words used.)

Form B (Type III)

Dated   Signed

                                        Designation

On behalf of the   Council

(Local planning authority for the area in which the proposed development is to be carried out)

Two completed copies of this Form, both signed, should be returned to the applicant for submission by them to the Department of Trade and Industry.

Appendix 14

Modifications to Overhead Line on Existing Route (including reconductoring replacement of supports or placing additional supports)

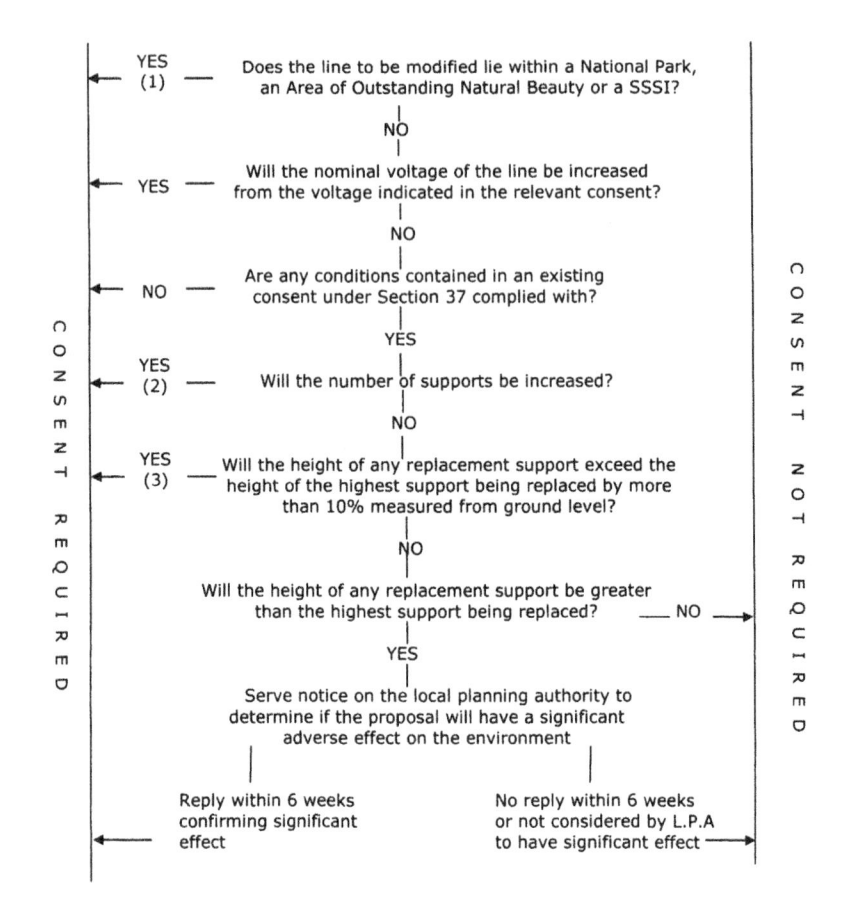

Electricity Wayleaves, Easements and Consents

CONSENT REQUIRED ← | → CONSENT NOT REQUIRED

- **YES (1)** ← Does the line to be modified lie within a National Park, an Area of Outstanding Natural Beauty or a SSSI?
  - NO ↓
- **YES** ← Will the nominal voltage of the line be increased from the voltage indicated in the relevant consent?
  - NO ↓
- **NO** ← Are any conditions contained in an existing consent under Section 37 complied with?
  - YES ↓
- **YES (2)** ← Will the number of supports be increased?
  - NO ↓
- **YES (3)** ← Will the height of any replacement support exceed the height of the highest support being replaced by more than 10% measured from ground level?
  - NO ↓
- Will the height of any replacement support be greater than the highest support being replaced? —— NO →
  - YES ↓
- Serve notice on the local planning authority to determine if the proposal will have a significant adverse effect on the environment
  - Reply within 6 weeks confirming significant effect ←
  - No reply within 6 weeks or not considered by L.P.A to have significant effect →

Notes
(1) A consent is necessary for only that part of the line to be modified which lies in a National Park, Area of Outstanding Natural Beauty or a SSSI.
(2) A consent is necessary for the additional support(s) only.
(3) A consent is necessary only for the support(s) which exceed the height of the highest support being replaced by more than 10%.

Appendix 15

Permanent Overhead Line Diversions

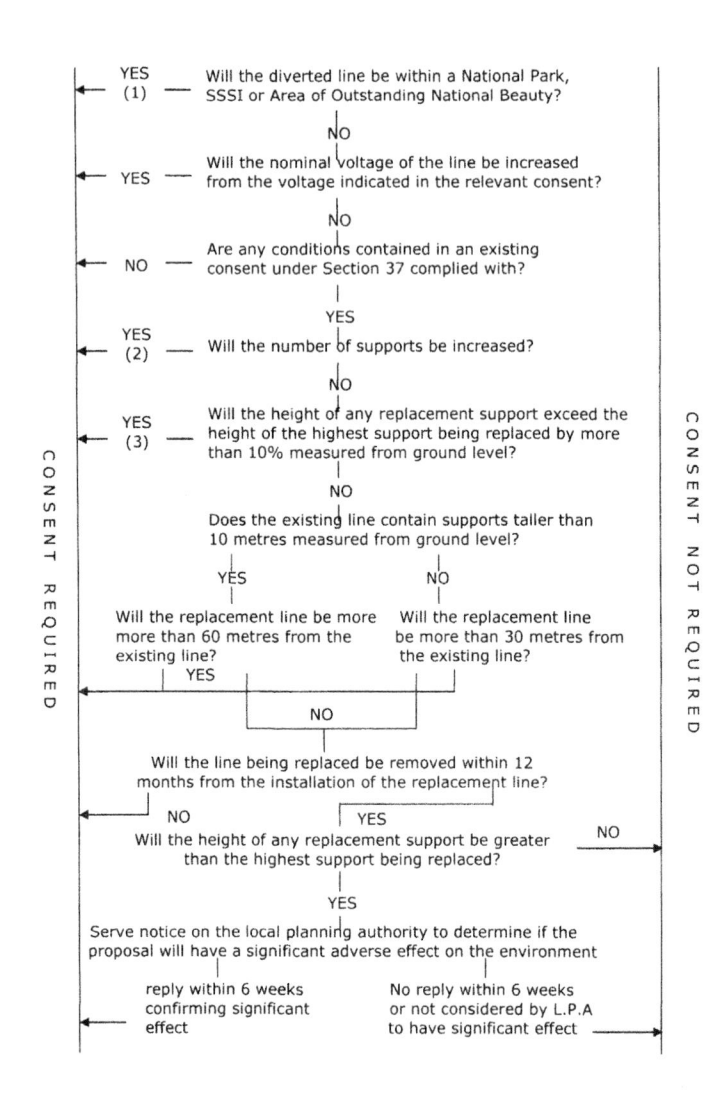

Electricity Wayleaves, Easements and Consents

```
                                                                            
          ┌── YES ──  Will the diverted line be within a National Park,
          │   (1)     SSSI or Area of Outstanding National Beauty?
          │                         │
          │                        NO
          │           Will the nominal voltage of the line be increased
          ├── YES ──  from the voltage indicated in the relevant consent?
          │                         │
          │                        NO
          │           Are any conditions contained in an existing
          ├── NO  ──  consent under Section 37 complied with?
          │                         │
          │                        YES
          │   YES
          ├── (2) ──  Will the number of supports be increased?
          │                         │
          │                        NO
          │           Will the height of any replacement support exceed the
          │   YES     height of the highest support being replaced by more
          ├── (3) ──  than 10% measured from ground level?
  C       │                         │                                   C
  O       │                        NO                                   O
  N       │           Does the existing line contain supports taller than  N
  S       │           10 metres measured from ground level?             S
  E       │                   │              │                          E
  N       │                  YES             NO                         N
  T       │                   │              │                          T
          │           Will the replacement   Will the replacement line   
  R       │           line be more           be more than 30 metres from  N
  E       │           than 60 metres from    the existing line?         O
  Q       │           the existing line?                                T
  U       │                │  YES                                       
  I       └────────────────┤                                            R
  R                        │                                            E
  E                        └──────── NO ────────┐                       Q
  D                                              │                      U
                                                 │                      I
                        Will the line being replaced be removed within 12  R
                        months from the installation of the replacement line?  E
                                │                                       D
          ┌───────────────── NO │      │ YES ──────────┐
                                                       │
                        Will the height of any replacement support be greater    NO
                        than the highest support being replaced?     ────────►
                                        │
                                       YES
                        Serve notice on the local planning authority to determine if the
                        proposal will have a significant adverse effect on the environment
                                        │
                          reply within 6 weeks    No reply within 6 weeks
                          confirming significant  or not considered by L.P.A
          ◄──────────────  effect                 to have significant effect ──────►
```

Notes
(1) A consent is necessary for only that part of the line to be modified which lies in a National Park, Area of Outstanding Natural Beauty or a SSSI.
(2) A consent is necessary for the additional support(s) only.
(3) A consent is necessary only for the support(s) which exceed the height of the highest support being replaced by more than 10%.

Appendix 16

Temporary Overhead Line Diversions

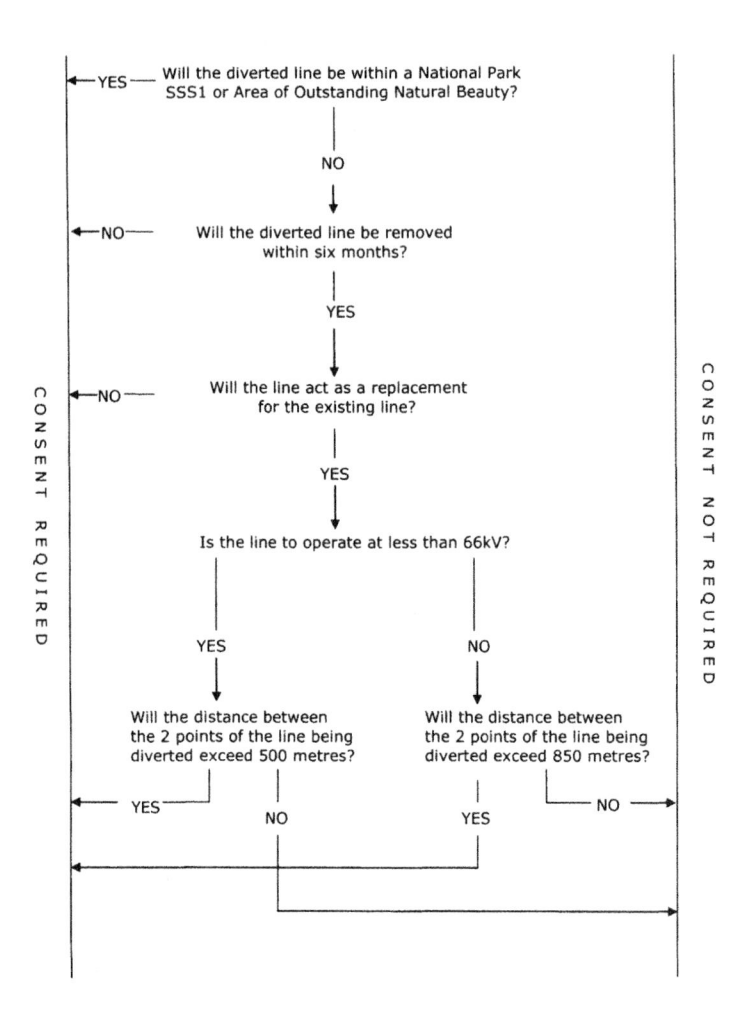

Electricity Wayleaves, Easements and Consents

```
                                                                                            C
  C                                                                                         O
  O      ←―YES――  Will the diverted line be within a National Park                          N
  N               SSS1 or Area of Outstanding Natural Beauty?                               S
  S                                  │                                                      E
  E                                  NO                                                     N
  N                                  ↓                                                      T
  T      ←―NO――   Will the diverted line be removed
                  within six months?                                                        N
  R                                  │                                                      O
  E                                  YES                                                    T
  Q                                  ↓
  U      ←―NO――   Will the line act as a replacement                                        R
  I               for the existing line?                                                    E
  R                                  │                                                      Q
  E                                  YES                                                    U
  D                                  ↓                                                      I
                                                                                            R
                  Is the line to operate at less than 66kV?                                 E
                          │                         │                                       D
                         YES                        NO
                          ↓                         ↓
                Will the distance between    Will the distance between
                the 2 points of the line     the 2 points of the line
                being diverted exceed        being diverted exceed
                500 metres?                  850 metres?
                     │                            │
              ―YES―┘  └―NO―┐           ┌―YES―┘   └――NO――→
                            │           │
```

300

Permitted Development by Licence Holders

Extract from Town and Country Planning (General Permitted Development Order) 1995 (Statutory Instrument 418/1995)

Schedule 2 Permitted Development
Part 17 Development by Statutory Undertakers

Class G Electricity undertakings

Permitted development

G. Development by statutory undertakers for the generation, transmission or supply of electricity for the purposes of their undertaking consisting of —

  (a) the installation or replacement in, on, over or under land of an electric line and the construction of shafts and tunnels and the installation or replacement of feeder or service pillars or transforming or switching stations or chambers reasonably necessary in connection with an electric line;

  (b) the installation or replacement of any telecommunications line which connects any part of an electric line to any electrical plant or building, and the installation or replacement of any support for any such line;

(c) the sinking of boreholes to ascertain the nature of the subsoil and the installation of any plant or machinery reasonably necessary in connection with such boreholes;

(d) the extension or alteration of buildings on operational land;

(e) the erection on operational land of the undertaking of a building solely for the protection of plant or machinery;

(f) any other development carried out in, on, over or under the operational land of the undertaking.

Development not permitted
G.1 Development is not permitted by Class G if—

(a) in the case of any Class G(a) development—

   (i) it would consist of or include the installation or replacement of an electric line to which section 37(1) of the Electricity Act 1989 (consent required for overhead lines) applies; or

   (ii) it would consist of or include the installation or replacement at or above ground level or under a highway used by vehicular traffic, of a chamber for housing apparatus and the chamber would exceed 29 cubic metres in capacity;

(b) in the case of any Class G(b) development—

   (i) the development would take place in a National Park, an area of outstanding natural beauty, or a site of special scientific interest;

   (ii) the height of any support would exceed 15 metres; or

   (iii) the telecommunications line would exceed 1,000 metres in length;

(c) in the case of any Class G(d) development—

   (i) the height of the original building would be exceeded;

(ii) the cubic content of the original building would be exceeded by more than 25% or, in the case of any building on article 1(5) land, by more than 10%, or

(iii) the floor space of the original building would be exceeded by more than 1,000 square metres or, in the case of any building on article 1(5) land, by more than 500 square metres;

(d) in the case of any Class G(e) development, the building would exceed 15 metres in height, or

(e) in the case of any Class G(f) development, it would consist of or include—

(i) the erection of a building, or the reconstruction or alteration of a building where its design or external appearance would be materially affected, or

(ii the installation or erection by way of addition or replacement of any plant or machinery exceeding 15 metres in height or the height of any plant or machinery replaced, whichever is the greater.

Conditions

G.2 Development is permitted by Class G subject to the following conditions—

(a) in the case of any Class G(a) development consisting of or including the replacement of an existing electric line, compliance with any conditions contained in a planning permission relating to the height, design or position of the existing electric line which are capable of being applied to the replacement line;

(b) in the case of any Class G(a) development consisting of or including the installation of a temporary electric line providing a diversion for an existing electric line, on the ending of the diversion or at the end of a period of six months from the completion of the installation (whichever is the sooner) the temporary electric line shall be removed and the land on which any operations have been carried out to install that line shall be restored as soon as reasonably practicable to its condition before the installation took place;

(c) in the case of any Class G(c) development, on the completion of that development, or at the end of a period of six months from the beginning of that development (whichever is the sooner) any plant or machinery installed shall be removed and the land shall be restored as soon as reasonably practicable to its condition before the development took place;

(d) in the case of any Class G(e) development, approval of details of the design and external appearance of the buildings shall be obtained, before development is begun, from—

   (i) in Greater London or a metropolitan county, the local planning authority,

   (ii) in a National Park, outside a metropolitan county, the county planning authority,

   (iii) in any other case, the district planning authority.

Interpretation of Class G

G.3 For the purposes of Class G(a), "electric line" has the meaning assigned to that term by section 64(1) of the Electricity Act 1989 (interpretation etc. of Part 1).

G.4 For the purposes of Class G(b), "electrical plant" has the meaning assigned to that term by the said section 64(1) and "telecommunications line" means a wire or cable (including its casing or coating) which forms part of a telecommunication apparatus within the meaning assigned to that term by paragraph 1 of Schedule 2 to the Telecommunications Act 1984 (the telecommunications code).

G.5 For the purposes of Class G(d), (e) and (f), the land of the holder of a licence under section 6(2) of the Electricity Act 1989 (licences authorising supply etc.) shall be treated as operational land if it would be operational land within section 263 of the Act (meaning of "operational land") if such licence-holders were statutory undertakers for the purpose of that section.

Appendix 18

Wayleave Payments

Wayleave Payments — 1 April 2005 to 31 March 2006

	Owner	Occupier			Owner/Occupier		
		Arable	Grassland	Hedgerow	Arable	Grassland	Hedgerow
Pole	£5.38	£11.37	£1.77	£0.70	£16.75	£7.15	£6.08
H Pole	£6.85	£13.45	£2.98	£1.57	£20.30	£9.83	£8.42
Stay	£1.31	£9.80	£2.09	£1.75	£11.11	£3.40	£3.06
Additional Stay	£1.31	£4.84	£1.04	£0.86	£6.15	£2.35	£2.17
Tower 1	£17.26	£24.40	£5.39	*	£41.66	£22.65	*
Tower 2	£19.94	£27.56	£7.55	*	£47.50	£27.49	*
Tower 3	£23.31	£29.67	£9.00	*	£52.98	£32.21	*
Tower 4	£31.19	£31.51	£10.26	*	£62.70	£41.45	*
Tower 5	£33.98	£36.26	£13.15	*	£70.24	£47.13	*
Tower 6	£36.72	£48.23	£14.42	*	£84.95	£51.14	*
Tower 7	£45.18	£48.23	£17.13	*	£93.41	£62.31	*
Tower 8	£54.89	£60.18	£20.01	*	£115.07	£74.90	*
Tower 9	£69.01	£60.18	£22.72	*	£129.19	£91.73	*
Tower 10	£77.98	£73.72	£25.43	*	£151.70	£103.41	*
Tower 11	£93.17	£73.72	£28.14	*	£166.89	£121.31	*
Tower 12	£101.16	£87.77	£31.03	*	£188.93	£132.19	*
Tower 13	£115.39	£87.77	£31.03	*	£203.16	£146.42	*
Lines only — poles	£1.08	£0.00	£0.00	£0.00	£1.08	£1.08	£1.08
Lines only — towers	£1.54	£0.00	£0.00	£0.00	£1.54	£1.54	£1.54
PB single — unstayed	£5.71	£11.37	£1.77	£0.70	£17.08	£7.48	£6.41
PB double — unstayed	£25.94	£21.54	£3.77	£3.14	£47.48	£29.71	£29.08
PB — multi stayed	£43.81	£83.58	£38.43	*	£127.39	£82.24	*

* Appropriate rate for adjacent fields applied

Electricity Wayleave Payments to Landowners — 1 April 2005 to 31 March 2009

		Existing 1.4.2004	1.4.2005	1.4.2006	1.4.2007	1.4.2008
Pole	n/a	£5.20	£5.38	£5.57	£5.77	£5.98
H Pole	n/a	£6.62	£6.85	£7.09	£7.35	£7.62
Stay	n/a	£1.27	£1.31	£1.36	£1.41	£1.46
Additional Stay	n/a	£1.27	£1.31	£1.36	£1.41	£1.46
Tower 1	2.6m^2	£16.68	£17.26	£17.86	£18.51	£19.19
Tower 2	3.8m^2	£19.27	£19.94	£20.64	£21.39	£22.17
Tower 3	4.6m^2	£22.52	£23.31	£24.13	£25.01	£25.92
Tower 4	5.3m^2	£30.14	£31.19	£32.28	£33.46	£34.68
Tower 5	6.9m^2	£32.83	£33.98	£35.17	£36.45	£37.78
Tower 6	7.6m^2	£35.48	£36.72	£38.00	£39.39	£40.82
Tower 7	9.1m^2	£43.65	£45.18	£46.76	£48.47	£50.24
Tower 8	10.7m^2	£53.03	£54.89	£56.81	£58.88	£61.03
Tower 9	12.2m^2	£66.68	£69.01	£71.43	£74.04	£76.74
Tower 10	13.7m^2	£75.34	£77.98	£80.71	£83.66	£86.71
Tower 11	15.2m^2	£90.02	£93.17	£96.43	£99.95	£103.60
Tower 12	16.8m^2	£97.74	£101.16	£104.70	£108.52	£112.48
Tower 13	18.3m^2	£111.49	£115.39	£119.43	£123.79	£128.31
Lines only — poles	n/a	£1.04	£1.08	£1.12	£1.16	£1.20
Lines only — towers	n/a	£1.49	£1.54	£1.59	£1.65	£1.71
PB single — unstayed	n/a	£5.52	£5.71	£5.91	£6.13	£6.35
PB double — unstayed	n/a	£25.06	£25.94	£26.85	£27.83	£28.85
PB — multi stayed	n/a	£42.33	£43.81	£45.34	£46.99	£48.70

Electricity Industry Payments to Occupiers Explanatory Notes

Electricity wayleave payments represent compensation for annual losses for interference on agricultural land caused by the presence of electricity lines in England and Wales. Commercial horticultural land has not been included in the assessment of compensation. The payments are reviewed by the Electricity Industry with the National Farmers Union, Country Land & Business Association and Farmers Union of Wales. The following guidelines will be used by licence-holders when applying wayleave payments.

Where a support obstructs agricultural operations the appropriate compensation rate will be applied.

Arable includes grass leys of up to and including five years duration and intensively managed grassland.

Permanent pasture is grassland on which agricultural operations take place but it excludes rough grazing and moorland.

The rate for an H pole assumes a gap between the two limbs at the pole centres up to and including 3 m.

The higher payment for a stay will apply to a single stay where the annexed pole is outside the cultivated area or where a stay is installed at more than 45 degrees to an adjacent stay.

Payments for additional stays will be applied where the adjacent stay is within 45 degrees of the initial stay.

Where an isolated stay causes a significant problem owing to its position in relation to the field boundary, then licence-holders will consider individual claims on their own merits.

Appendix 21

Tower Size Allocations for Wayleave Payments

Measurements of towers will be taken from the exterior angle of concrete bases at ground level. If part of a tower extends into a field then a pro-rata payment will be applied. For each tower category the following land area has been used to calculate compensation:

Tower Number	Tower Size
Tower 1	2.6 m^2
Tower 2	3.8 m^2
Tower 3	4.6 m^2
Tower 4	5.3 m^2
Tower 5	6.9 m^2
Tower 6	7.6 m^2
Tower 7	9.1 m^2
Tower 8	10.7 m^2
Tower 9	12.2 m^2
Tower 10	13.7 m^2
Tower 11	15.2 m^2
Tower 12	16.8 m^2
Tower 13	18.3 m^2

Appendix 22

Lands Tribunal   Form R

Notice of Reference Form

- First unfold the form and read the accompanying notes on pages 3 & 4.
- Complete the form in **BLOCK CAPITALS** and in **BLACK INK** or preferably typed.

Part A — Names of Parties
Name of Claimant..
See note 1.1 opposite
Address..
....
....
Telephone No....... Fax No .... Status ....

Name of Claimant's Solicitor or other Representative ....
See note 1. 2 opposite
Address..
....
....
Telephone No....... Fax No .... Status ....

Name of Compensating/Acquiring Authority......................
....
See note 1.3 opposite
Address...
....
....
Telephone No .... Fax No .... Status ....

Name of Compensating/Acquiring Authority's
Solicitor or other Representative.................................
See note 1. 4 opposite
Address...
....
....
Telephone No .... Fax No .... Status ....

Part B — The Land / Property
Description of land/property ....
See note 2 opposite
....
....
Location or Address of land/property ....
....
....

Part C — The Claim
Nature of interest claimed for ....
....
....
See note 3 opposite
Statutory provision (indicate the relevant Act and section) or agreement under which the reference is made....................
....

Approximate amount of the claim £...............................
See note A opposite

This is a: (tick one box only)
  claim for compensation following a compulsory purchase order

claim for compensation following acceptance of a purchase notice or blight notice
other claim for land compensation
reference by consent: (*indicate nature of dispute*) ....
See note B opposite ....
....
....
....

Where compensation is claimed for compulsory purchase, has the acquiring authority entered upon the land or possession been given to the authority?
      Yes   No
If yes, on what date?   .... 19/20   ....

Expert Witnesses *See note C opposite*
(*tick one box on each line*)
C(i) Do you intend to call an expert witness at the hearing of the reference?   Yes   No

C(ii) Do you intend to apply for permission to call more than one expert witness at the hearing of the reference?   Yes   No

Part D — Type of Procedure Arrangements *See note D opposite*
(*tick one box only*)
Which type of procedure do you wish to be used in this reference?
Standard procedure   Special procedure   Simplified procedure   Written representation procedure

Part E —- Signature And Date *See note E opposite*
I refer this dispute to the Lands Tribunal for decision. I am:
(*tick one box only*)

    the claimant
    the solicitor for the claimant
    the agent for the claimant and I enclose a signed authority to act
    the authorised officer of the acquiring/compensating authority
    the solicitor for the acquiring/compensating authority

and I accept responsibility for the conduct of the case and the payment of fees.
I have paid the setting-down fee of £50.

Signed................   Date............   *See note F opposite*

Name in capitals ....

Enclosures: Fee, copies of Notice to Treat, Notice of Entry, Notice of Claim (and any amendments).

LANDS TRIBUNAL　　　　　　　　　　　　　Form R
Notice of Reference Form

This form is to be used for any **reference** under Rules 9–11 of the Lands Tribunal Rules 1996 (except Blight Notice and counter-notice cases under section 151 of Town & Country Planning Act 1990 for which use Form **BNO**)

Use of this form is not obligatory; it is intended to help users meet the requirements of rule 10. The Lands Tribunal green Explanatory Leaflet may also be helpful

First open out the form and read the notes on pages 3 and 4, then complete the form on pages 1 and 2

Note A VALUE OF THE CLAIM
Give here the approximate value you place on the claim (i.e. the amount you are provisionally contending for).
Do not give confidential, without prejudice or sealed offer figures.

Note B REFERENCES BY CONSENT
If this is a reference by consent, indicate briefly the nature of the dispute, and attach a copy of the agreement signed by both parties providing for disputes to be referred to the Lands Tribunal.

Note C EXPERT WITNESS(ES)
An expert witness is a witness who will give opinion evidence in support of your case and who may have a relevant professional

Lands Tribunal Form R

qualification. Tick the appropriate box in each case. If you intend to call an expert witness, the Tribunal will ask you in due course to lodge documents about the evidence to be given (reports, valuations, etc.).
If the claim relates to mineral valuations or business disturbance you may call two expert witnesses. In any other case, leave of the Tribunal will be required to call more than one expert witness. If you indicate that you wish to call two or more expert witnesses, the Tribunal will write to you about this. The same provisions apply to all parties.

Note D TYPE OF PROCEDURE ARRANGEMENTS
Most cases follow the **Standard Procedure**; if you think your reference should follow the **Simplified Procedure**, the **Special Procedure**, or the **Written Representations Procedure** tick the box and the Tribunal will contact you about it. See the Lands Tribunal Explanatory Leaflet for more information about the procedures.

Note E SIGNATURE AND DATE
Tick the appropriate box. A claimant in person, a solicitor instructed to act, or an officer employed by the appropriate Authority may sign the form. However an agent other than a solicitor may not sign unless there is attached to the form a written authority to act signed by the client.

Note F ENCLOSURES
A fee of £50 is payable on lodging each notice of reference and must be enclosed. (Please make cheques or postal orders payable to HM Paymaster General). A copy of the notice to treat (if any), notice of entry (if any) and of any notice of claim or amended notice of claim must be enclosed.

DELIVERY OF NOTICE OF REFERENCE & DOCUMENTS TO THE TRIBUNAL
The completed Notice of Reference together with enclosures (see Note F) must be sent to:

The Registrar, **Lands Tribunal,** **Procession House,** **55 Ludgate Hill,** **London EC4M 7JW**	**DX: 149065 Ludgate Hill 2** **Tel: 020 7029 9780** **Fax: 020 7029 9781**

LANDS TRIBUNAL Form R
Notice of Reference Form
Please read the notes below and overleaf

Note 1.1 Claimant
Give the full name and usual postal address of the claimant i.e. the person or persons claiming compensation in respect of the land or property described in part B. The status of the claimant may be an individual, limited company, plc, partnership, firm or corporation, or two or more of these acting jointly; please indicate.

Note 1.2 Claimant's Representative
If representatives are instructed to act, please indicate their name, address, telephone number, fax number and status (e.g. Solicitors, Chartered Surveyors, consultant etc.). Please note that all correspondence and documents including hearing notifications will be sent to the representatives, not direct to the claimant. If a representative ceases to act the claimant (or a solicitor, if acting) must inform the Tribunal at once.

Note 1.3 Compensating/Acquiring Authority
Give the full name and usual postal address of the authority against which the claim for compensation is made, or which is liable for the payment of compensation.

Note 1.4 Compensating/Acquiring Authority's Representative
Give the full name and address, telephone number and status (e.g. Solicitor, Chartered Surveyor, Consultant, etc.) of the Authority's representative, if any.

Note 2 The Land/Property
Description of land: give particulars of the land or property which is the subject of this reference e.g. dwellinghouse, shop, factory, agricultural land, vacant land, car park, etc. The approximate area of the land in sq. metres or hectares may be given if relevant.

Location of land: give the full postal address. If the land cannot be identified by a postal address e.g. agricultural land, please give O. S. map references showing the location of the land, or attach a copy map suitably marked.

Note 3 Nature of Claim
Interest: indicate the nature of the interest claimed for (e.g. freehold, leasehold, business tenancy).

Statutory provision: state the short title of the Act of Parliament and the relevant sections(s) of the Act which give the right to claim compensation and/or refer the disputed claim to the Lands Tribunal. If this is a Reference by Consent, indicate the Agreement providing for disputes to be referred to the Tribunal.

Notes on completing Form R

[The abbreviations in the footnotes are explained in Chapter 16]

A.23.1 Claimant's details

Form R is drawn on the basis that it will always be the landowner/occupier who is the claimant. It is essential that where the claimant is an individual, the full name and postal address is given. Where the claimant is a company or partnership then the full official business name must be given together with the registered office address.

A.23.2 Respondent's details

On Form R the respondent to the reference is described as the "Compensating/Acquiring Authority". This is because the majority of cases involving a compensation reference arise out of compulsory purchase. However, it is correct to put here the details of the licence-holder to whom the necessary or the non-statutory[1] wayleave has been granted. If more then one person has an interest in the land, then all those who have such an interest must be named.

---

1   For discussion of these terms see Chapters 6, 7 and 8.

A.23.3 Representatives

The claimant must also name his representative giving full contact details. With one exception, there is no restriction on who may be a representative. A claimant may appear in person or be represented by counsel or solicitors or by any other person with the leave of the Tribunal or if it is an interlocutory hearing,[2] the leave of the President or the registrar.[3] The exception is that a valuation officer who is a party may not appear in person without permission.

The practice directions expand this provision. A legal executive who is a Fellow of the Institute of Legal Executives and has a certificate covering the Lands Tribunal will be granted permission to represent his or her client on application made at or prior to the hearing. An application for permission for a friend to represent a party who is an individual or one spouse to represent the other will readily be granted and may be made at the hearing. Otherwise, applications for permission to represent a party should be made in good time prior to the hearing, but the Tribunal may grant permission at the hearing as a matter of discretion.[4]

So far as expert witnesses are concerned, in simple cases permission will usually be granted for a surveyor or valuer to represent a party in order to avoid the additional costs of separate representation. In cases allocated to the simplified procedure, such representation may well be the norm. However, the Lands Tribunal is hesitant about an expert acting both as an advocate and a witness in other cases. In general, the Lands Tribunal regard it as difficult and undesirable for the same person to act both as advocate and expert witness. Accordingly permission will not be granted for a non-lawyer to represent a party in any case where the Tribunal considers that the responsibilities of advocate and of expert witnesses are likely to conflict.[5]

There are additional provisions relating to solicitors. Where a solicitor commences or responds to a reference he should be noted on the record at the Tribunal.[6] A party who began acting in person can

---

2 That is to say an application made by a party in the course of proceedings pending or ancillary to the final determination of the case.
3 1996 R37(1).
4 PD17.1
5 PD17.1.
6 1996 R53(1).

appoint a solicitor at any time to act for him.[7] A party can at any time appoint alternative solicitors to act for him in place of those previously on the record or continue to represent himself in person.[8] Wherever there is a change of any kind, notices confirming the change must be served on all relevant parties including the outgoing solicitor and those noticed must also be filed with the Lands Tribunal.[9] It is vitally important that the notices are filed with the Lands Tribunal because whichever solicitor is on the record is liable for the fees payable to the Lands Tribunal.[10] Note that this is a personal liability on the part of the solicitor to pay the fees due from the party whom he represents.

A.23.4 Description of land

The next task is to identify the land both by description and by location.[11] The authors recommend that a description be given which follows the relevant title document, for example : "the land known as [as set out in the Property Registry if the title is registered] and comprised within registered title number [   ]." Or "the land known as [   ] as more particularly described in the conveyance dated [   ] and made between [   ]" (for registered title). Further if an address is given, this should be the full postal address including postcodes. Location should, so far as possible have grid references relating to an ordnance survey plan as requested by the Lands Tribunal. It is permissible to attach a plan. The approximate area of the land in square metres or hectares may also be given if relevant which, in the majority of cases it will be.

A.23.5 Identifying relevant legal interests

Form R then asks the claimant to state the nature of the interests claimed.[12] This is primarily a reference to the legal interest which has to be valued on basis that it has been injuriously affected by the

---

7   1996 R53(2).
8   1996 R53(3).
9   1996 R53(4) and 53(2) and 53(3).
10   1996 R53(5) and PD21.
11   1996 R10(3)(c) — note the Rule requires only the address or the description. However, see the discussion in the text above.
12   1996 R10(3)(d).

necessary or voluntary wayleave. The usual candidates will be freehold/leasehold business tenancies. However, any legal or equitable interest[13] will qualify. Note that the form at this point will need adaptation if the claim is for damage to moveables under paragraph 7(3) of schedule 4 to the Ac.t[14]

A.23.6 Jurisdiction

The Claimant must then indicate the statutory provision or the agreement under which the reference is made.[15] For our purposes this will (subject to the points made in Chapter 16, section 16.7), usually be paragraph 7(2) or 7(3) of Schedule 4 to the Act or pursuant to the Arbitration Act 1996 under a provision in an express or implied wayleave. Remember that the Lands Tribunal treats the fact that both the parties to the reference have signed an agreement providing for reference to the Lands Tribunal as a sufficient consent for the purpose of section 1(5) of the 1949 Act.

If the agreement is an implied wayleave[16] then the authors recommend that a covering letter be sent explaining how the intended parties to the reference are bound by the implied wayleave and any consent forming part of its terms. Any dispute as to whether the parties are so bound will be an obvious candidate for the trial of preliminary issue.[17]

A.23.7 Amount at stake

The next requirement is that the claimant must identify the approximate amount of the claim. This is an official statement of the amount for which the claimant is provisionally contending. Care must be taken at this point. If the amount is exaggerated, that can have the effect of reducing the claimant's entitlement to costs.[18] On the other hand the claimant is not required to reveal any negotiating position he has discussed with his adviser. Furthermore, the Lands Tribunal must

---

13   See Chapter 5 for discussion of the term "legal or equitable interest".
14   In which case the area of the land will presumably not be relevant to the claim.
15   1996 R10(e).
16   See Chapter 5.
17   See Chapter 16 for further discussion of preliminary issues.
18   See Chapter 16 for further discussion of the principles on which costs are awarded.

not on any account be told of any actual offers or counter-offers made. Hence the notes to Form R warn the claimant *not* to give confidential without prejudice or sealed offer figures. Submisions regarding section 4 of the Land Compensation Act 1961 should be made in the Particulars of Claim.

A.23.8 Summary of the claim

The next section of Form R asks the claimant to summarise which type of claim the Lands Tribunal is being asked to determine. The choices so far as this book is concerned are: "Other claim for land compensation" or "a reference by consent". This would include a case in which a clause in the wayleave authorising a reference is being implemented. In that case, a brief statement explaining that the circumstances have arisen which have brought the clause into operation must also be made. For example, the value of the land has diminished by reason of the existence of the electric line. "References by consent" will also include an agreed submission of the question of compensation following the voluntary grant of an easement (albeit such grant may have been made under the shadow of compulsion).

A.23.9 Expert witnesses

The claimant must then indicate his position regarding expert witnesses. An expert witness is a witness who gives evidence of his opinion. He does not give evidence of fact, ie he does not testify to what he actually saw, heard or did. The expert has to be qualified by training and experience to give and express the opinion in question.[19] Where the expert is a valuer he will obviously have a recognised professional qualification. However it is not automatically required that an expert should be so professionally qualified. Whoever the expert is, he must nevertheless comply with regulations regarding experts. Unless the Lands Tribunal agrees otherwise, only one expert witness may be called.[20] There are exceptions for mineral valuations and business disturbance.[21] If you intend to apply for permission to

---

19   PD16.1.
20   1996 R42(2).
21   1996 R42(3).

call more than one expert at the hearing, you must say so. How one goes about making such an application and on what grounds is discussed below.

A.23.10 Procedure

The claimant then has to express a preference for a procedure. As indicated in Chapter 16 there are four possibilities.[22]

Finally, there are formal requirements for identification, date and signature.[23]

A.23.11 Attachments to the notice of reference

The only document to be attached to the notice of reference should be the wayleave and any plan.

A.23.12 Formalities

Note that for our purposes the standard enclosures listed on Form R[24] are all related to compulsory purchase under the Compulsory Purchase Act 1965 and are therefore irrelevant to the compensation references with which we are concerned.

A.23.13 Fees

A fee is payable. The current schedule of fees is obtainable from the website or from the offices of the Lands Tribunal.

---

22 They are: standard procedure, special procedure, simplied procedure, written representations.
23 1996 R10(f).
24 Notice to treat, notice of entry, notice of claim.

Appendix 24

Notice Periods and Dates of Service

A.24.1 Overview of this appendix

On a number of occasions it will be important for those dealing with wayleaves and consents to calculate first the date on which a document has been served and second, the date on which a period after service expires. Examples of periods and dates which will be relevant to the matters considered in this book include:

- notices to terminate a non-statutory wayleave
- notices to remove electric lines
- time-limits for the removal of electric lines
- time-limits for the service of required documents in Lands Tribunal proceedings.

In this appendix we consider the meaning of service and also set out the various sources of information and obligations contained within the relevant statutory materials. We also provide some commonly accepted guidelines for calculating notice periods and dates of service.

A.24.2 The purpose and meaning of "service"

"The purpose of serving a document is to ensure that its contents are available to the recipient".[1] In some cases the very fact that a document has been made available is enough to create legal consequences. An example is the service of a notice to terminate non-statutory wayleaves. Provided the licence-holder has the required amount of warning (or notice) of the intended termination date, the making available of the document has that necessary legal consequence. The wayleave therefore terminates at the expiration of the relevant notice period.

Further, service of the document can be designed to require the recipient to respond to the document or take some other action within a specified period, failing which the recipient is responsible for the consequences of not doing so. This is particularly important in the context of any form of court or other judicial application. "The function of service is primarily to bring to the attention of the person to be served the fact that he is being sued".[2] "It is a fundamental principle of our law that no one is to be found guilty or made liable by an order of any tribunal unless he has been given fair notice of the proceedings so as to enable him to appear and defend them".[3]

Whether one talks about a notice being given or sent or about service being effected, there is in essence a process which has a number of distinct stages. First the sender takes steps to have the notice or other document delivered to the recipient. In many cases this will be by post. In some cases it will be by handing the notice or other document personally to the recipient (personal service) or it will be by leaving the document at an official address of the recipient at which he is taken by the law to be present, such that the sender can be confident the document will come to the recipient's attention. In some cases the law will deem a recipient to have received a document provided a proper process has been taken. An example of this is the case of *In Re 88 Berkeley Road, Rickwood v Turnsek* [1971] to WLR 307 where the recipient was deemed to have received a notice sent by recorded delivery, even

---

1 *Hastie & Jenkerson v McMahon* [1990] 1 WLR 1990 at 1575 *per* Woolf LJ (as he then was). This was the Court of Appeal judgment which permitted service of court documents by fax.
2 *White v Weston* [1968] 2 QB 647, *per* Russell LJ at 658.
3 *Ex parte Rossi* [1956] 1 QB 682, *per* Denning LJ (as he then was) at 691.

though the recorded delivery was signed for, not by the recipient but rather by the co-resident of the address to which the letter had been sent. The intended recipient was, therefore, bound by the notice, even though as a matter of fact, she had not seen it. The decision may seem all the more surprising when it is noted that the co-resident who signed was the person who sent the notice in the first place.

Consequently, although the process of service involves the sending or delivery of a document, actual receipt is not necessarily a requisite component. Deemed receipt may be sufficient for the purposes of enabling the sender to take advantage of the fact that the document has been "served". In connection with the service of court proceedings, for example, (which for these purposes would include proceedings for the Lands Tribunal) it will be sufficient if the person serving the documents has chosen an appropriate method as provided for by the Civil Procedure Rules or the 1996 Rules (whichever applies) and has correctly identified an address for service in accordance with those rules. It is only if the sender chooses to adopt an alternative method that the sender must then prove actual receipt, rather than take advantage of the deeming provisions.

A.24.3 Ordinary course of post

Section 7 of the Interpretation Act 1978 reads as follows:

> Where an Act authorises or requires a document to be served by post (whether the expression "served" or the expression "given" or "send" or any other expression is used) then, unless the contrary intention appears, the service is deemed to be effected by properly addressing, pre-paying and posting a letter containing the document and, unless the contrary is proved, to have been effected at the time at which the letter would be delivered in the ordinary course of post.

There is no further definition of ordinary course of post. There has been some judicial comment on that phrase. It covers both ordinary post and registered post.[4] It is for the person claiming that a document has been served in the ordinary course of post to prove what the ordinary course of post actually is in the relevant locality.[5] In this

---

4 *TO Supplies (London) Ltd* v *Jerry Creighton Ltd* [1952] 1 KB 42.
5 *Kemp* v *Wanklyn* [1894] 1 QB at 265.

connection, it is worth noting that those wishing to effect postal service for the purposes of the Civil Procedure Rules 1998 are now to use "ordinary first class post and Royal Mail Special Delivery".[6] The justification is that Royal Mail Special Delivery guarantees service the next day. However, it should be noted that for the purposes of calculating the date of deemed service by first class post, the Civil Procedure Rules assume that it is the second day after it was posted.

Finally, before relying on section 7 of the Interpretation Act 1978, it should also be noted that there will be circumstances when the statute has been excluded by some other provision. An example of this is *Beanby Estates Ltd v Egg Stores (Stamford Hill) Ltd* [2003] 1 WLR which held that formal notices served under the Landlord and Tenant Act 1954, section 25, were not covered by section 7 of the Interpretation Act 1978. Consequently, a notice under that section served through the post by recorded delivery to the addressee at his place of abode was irrebuttably deemed to have been served, not when it was received or deemed to have been received by the recipient, but rather when it was put in the post.

A.24.4 Calculating notice periods and dates for service

If a provision (for instance in a non-statutory wayleave) requires "not less than" a number of months notice before a certain event can take place, then the sender must ensure that there is deemed or actual service which is at least as far ahead of the intended date as the number of months specified. In calculating periods of time, the "corresponding date rule" is a helpful guide. This provides that when calculating time by reference to the calendar month, it is the number of the day which is important rather than an exact period of weeks. Thus, 30 September is 6 months from 30 March. 31 December is 6 months from 30 June.

If a provision in a document or a rule of Court uses the word "clear day" that means that the amount of notice that has to be given excludes the day of service and excludes the date on or by which the act or event to which the notice refers has to take place.

---

6   *Petford* v *Saw* [2002] 11 CL 53.

A.24.5 Section 109 of the Act reads as follows:

(1) Any document required or authorised by virtue of this Act to be served on any person may be served:-

   (a) by delivering it to him or by leaving it as his proper address or by sending it by post to him at that address; or
   (b) if the person is a body corporate, by serving it in accordance with paragraph (a) above on the Secretary of that body; or
   (c) if the person is a partnership, by serving it in accordance with Paragraph (a) above on a partner or person having control or management of partnership business.

(2) For the purposes of this section and section 7 of the Interpretation Act 1978 (which relates to documents served by post in its application to this section, the proper address of any person on whom a document is to be served shall be his last known address, except that:

   (a) in the case of service on a body corporate or its Secretary, it shall be the address of the registered or principal office of the body;
   (b) in the case of service on a partnership or a partner or a person having the control or management of a partnership business, it shall be the address or the principal office of the partnership;

   And for the purposes of this sub-section the principal office of a company registered outside the United Kingdom or of a partnership carry on business outside the United Kingdom, is its principal office within the United Kingdom.

(3) If a person to be served by virtue of this Act with any document by another has specified to that other an address within the United Kingdom, other than his proper address (as determined in pursuance of sub-section (2) above) as the one at which he or someone on his behalf will accept documents of the same description as that document, that address shall also be treated as his proper address for the purposes of this section and for the purposes of the said section 7 in its application to this section.

(4) If the name or address of any owner or occupier of land on whom by virtue of this Act any document is to be served can not after reasonable enquiry be ascertained, the document may be served by:-

   (a) addressing it to him by the description of "owner" or "occupier" of the land (describing it); and

(b) either leaving it in the hands of a person who is or appears to be resident or employed on the land or leaving it conspicuously affixed to some building or object on or near the land.

(5) This section shall not apply to any document in relation to the service of which provision is made by rules of Court.

(6) In this section "the Secretary" in relation to a local authority within the meaning of the Local Government Act 1972 ... means the proper officer within the meaning of that Act.

A.24.6 Section 196 of the Law of Property Act 1925 reads as follows:

(as amended by the Postal Services Act 2000)

(1) Any notice required or authorised to be served or given by this Act[7] shall be in writing.

(2) Any notice required or authorised by this Act to be served on a lessee or mortgagor shall be sufficient, although only addressed to the lessee or mortgagor by that designation, without his name, or generally to the persons interested, without any name, and notwithstanding that any person to be affected by the notice is absent under disability, unborn, or unascertained.

(3) Any notice required or authorised by this Act to be served shall be sufficiently served if it is left at the last-known place of abode or business in the United Kingdom of the lessee, lessor, mortgagee, mortgagor, or other person to be served, or, in the case of a notice required or authorised to be served on a lessee or mortgagor, is affixed or left for him on the land or any house or building comprised in the lease or mortgage, or, in the case of a mining lease, is left for the lessee at the office or counting house of the mine.

(4) Any notice required or authorised by this Act to be served shall also be sufficiently served, if it is sent by post in a registered letter addressed to the lessee, lessor, mortgagee, mortgagor, or other person to be served, by name, at the aforesaid place of abode or business, office or counting house, and if that letter is not returned by the postal operator

---

7 That is to say the Law of Property Act 1925.

(within the meaning of the Postal Services Act 2000) concerned undelivered; and that service shall be deemed to be made of time at which the registered letter would, in the ordinary course, be delivered.

(5) The provisions of this section shall extend to notices required to be served by any instrument affecting property executed or coming into operation after the commencement of this Act unless the contrary intention appears.

(6) This section does not apply to notices served in proceedings in the Court.

A.24.7 Civil Procedure Rules 1998

The relevant provisions from the Civil Procedure Rules would not strictly apply to any proceedings before the Lands Tribunal although, subject to any explicit provision of the 1996 Rules, they are a very helpful guide. They clearly would apply to any proceedings relating to the enforcement of rights and remedies. These provisions are to be found in Part 6 of the Civil Procedure Rules. These change with greater frequency than the statutory provisions and so they are not set out in full in this appendix. The most current form of any of the Civil Procedure Rules can always be obtained by going onto the website of the Ministry of Justice at *www.justice.gov.uk*. From the home page you need to click on the link headed "Guidance".

Relevant Acts of Parliament

Year	Title of the Act	Cross-Reference to section numbering in the text
1882	Electricity Lighting	2.2
1899	Electric Lighting (Clauses)	12.6, 13.14
1925	Law of Property	5.6
1947	Electricity	4.2
1949	Lands Tribunal	16.3
1954	Electricity Reorganisation (Scotland)	2.2
1957	Electricity	2.2
1961	Land Compensation	11.4, 14.6, 15.2, 15.3, 15.5, 16.9, 16.14.2, 16.14.13, 16.16, 16.21.2, 16.22, 16.23
1962	Pipelines	22
1965	Compulsory Purchase	11.3–11.5
1966	Mines (Working Facilities and Support)	8.4
1973	Land Compensation	11.3–11.5, 15.2, 15.3, 15.6, 15.8, 16.5
1978	Interpretation	4.6, 8.3
1980	Local Government Planning and Land	Footnotes to 16.17
1981	Acquisition of Land	11.3–11.5
1981	Supreme Court	17.3
1984	Telecommunications	15.18
1984	County Court	17.3

Year	Title of the Act	Cross-Reference to section numbering in the text
1985	Housing	15.19
1986	Insolvency	18.4
1989	Electricity	Overview in 4.1–4.6 inclusive and otherwise quoted extensively throughout the book. Chapters 8 to 10 inclusive examine Schedule 4 to the Act. Chapter 11 examines Schedule 3 to the Act. Chapter 12 considers applications for a Ministerial Consent under section 37 of the Act and Chapter 13 consider Public Enquiries held in connection with such applications.
1990	Town and Country Planning	12.2 and Footnotes to 15.19
1991	Water Industry	22
1992	Competition and Service (Utilities)	4.2, 4.4
1994	Telecommunications	14.4
1996	Arbitration	16.7
1997	Civil Procedure	Footnotes to 16.11.1 and extensively throughout Chapters 16–21 and Appendix 24 inclusive.
1998	Competition	4.2
1998	Late Payment of Commercial Debts (Interest)	17.3
2000	Utilities	4.2, 4.4, 6.9, 7.4
2002	Enterprise	4.2
2003	Sustainable Energy	4.2
2004	Energy	4.2

Appendix 26

Relevant Statutory Instruments

Year	Title of Regulation Rules or Order	Cross-reference to section numbering in the text
1967	Electricity (Compulsory Wayleaves) Hearings Procedure	10.3
1986	Insolvency	18.4
1988	Electricity Supply	3.2
	Town and Country Planning (Assessment of Environmental Effects)	12.3
1989	Electricity at Work	3.2
1990	Electricity Act 1989 (Consequential Modifications of Subordinate Legislation)	10.3
1990	Electricity (Application for Consent)	12.3
1990	Overhead Line (Exemption)	12.2, 12.5
1990	Electricity and Pipe-line Works (Assessment of Environmental Effect)	13.3
1990	Electricity Generating Stations and Overhead Lines (Inquiries Procedure)	13.1, 13.4, 13.6
1990	Fees for Inquiries (Standard Daily Amount)	13.15
1993	Judgment Debts (Rate of Interest)	18.7
1994	Town and Country Planning (Assessment of Environmental Effects) (Amendment)	12.3

Year	Title of Regulation Rules or Order	Cross-reference to section numbering in the text
1994	Electricity and Pipe-line Works (Assessment of Environmental Effect) (Amendment)	13.6
1995	Town and Country Planning (General Permitted Development)	12.2
	Acquisition of Land (Rate of Interest after Entry)	18.6
1996	Lands Tribunal	15.2 and extensively in Chapter 16
1996	Lands Tribunal (Fees)	13.13.11
1997	Electricity Supply Industry Fee Scales	15.20
1997	Electricity Generating Stations and Overhead Lines and Pipe-lines (Inquiries Procedures) (Amendment)	13.4, 13.6
1998	Civil Procedure	Chapter 16 and Part 5 extensively
2000	Electricity Works (Environmental Assessment) (England and Wales)	12.3
2000	Civil Procedure (Modification of Enactment)	16
2002	Electricity Safety Quality and Continuity	3.2
2006	Electricity Safety Quality and Continuity (Amendment)	Footnotes to 3.2

Appendix 27

Comparison of Utility Legislation in Respect of Bases for Compensation

Utility	Source of Statutory Powers	Need for Owners Consent	Basis of Compensation
Water and sewerage	Water Act 1991	By notice only	Loss in land value
Gas	Gas Act 1986	Voluntarily	Proportion of land value
Telecommunications	Telecommunications Act 1984	Voluntarily or imposed by Court	Loss in land value and consideration of benefit to user
Electricity	Electricity Act 1989	Voluntarily or following wayleave hearing or by compulsory purchase order	Agricultural values for wayleaves or loss in land value for permanent rights

Useful Addresses

Association of Electricity Producers
First Floor
17 Waterloo Place
London
SW1Y 4AR

Energy Networks Association
18 Stanhope Place
Marble Arch
London
W2 2HH

Lands Tribunal
Procession House
55 Ludgate Hill
London
EC4M 7JW

Department of Trade and Industry:
Wayleaves Manager
Licensing and Consents Unit
Department of Trade and Industry
1 Victoria Street
London SW1H 6ET
Tel: 020 7215 2742
Fax: 020 7215 2601

Health and Safety Executive
Rose Court
2 Southwark Bridge
London
SE1 9HS

Index

ancillary rights
    as part of the statutory menu......................................82
    implied in necessary wayleaves...................................67
    relevance generally 47
    remedy for breach of 223
application to Secretary of State for necessary wayleave
    form of application
        additional requirements......................................78
        contents...78
        pitfalls..79
        time periods..80
    procedure
        circumstances for making application 51, 65, 73, 74, 87
        DTI guidance 74
        notices
            notice to remove..76
            notice to terminate..75
            one notice procedure......................................76
            two notice procedure......................................76
        request for wayleave from licence-holder.........................77
        rights of entry onto land
            environmental issues............................29, 71, 82, 123
            lopping and felling of trees..................................82
            surveys/inspection 84
        Secretary of State's jurisdiction.................................81

compulsory purchase order
    compulsory acquisition between licence holders 106

343

```
compensation .................................................. 109
legislation .................................................... 105
procedure
    Acquisition of Land Act 1981 ................................ 107
    application to Lands Tribunal ............................... 107
    compensation code .......................................... 108
    whole of property .......................................... 109
usefulness of ................................................. 109
consents ........................................................ 113

dwellings
    Secretary of State: restriction on powers in respect of
        compulsory purchase ..................................... 110
        necessary wayleave ....................................... 81

easement
    alternative to wayleave ................................. 43, 70, 143
    benefit to landowner of granting ............................. 46
    benefit to licence-holder of taking .......................... 47
    compensation framework
        compensation code ..................................... 145, 148
        date of installation ..................................... 145
        principles of valuation
            blight and sterilisation ............................. 154
            capitalisation ....................................... 153
            development clause ................................... 152
            development site ..................................... 150
            disturbance .......................................... 149
            fibre-optics ......................................... 154
            injurious affection .................................. 149
            land taken ........................................... 148
            legal costs .......................................... 156
            lift and shift clause ................................ 152
            mineral extraction ................................... 151
            surveyors fees ....................................... 155
            tree planting and cutting ............................ 151
        time ............................................... 37, 70, 72
    contrasted with other rights .................................. 42
    definition ............................................... 37, 43
electric lines
    definition .................................................. 30
    overhead lines
        considerations when planning to put up a new electric line ........ 119
electricity industry
```

Index

current arrangements 12
current industry regulation
    licence conditions 29
    offences 28
    the regulator 28
    transfer of licence 28
DNO 11
history of 9, 16
infrastructure
    clearance to buildings and obstacles 20
    folded steel plate 18
    foundations 21
    P B structures 19
    spacing between supports 19
    steel lattice towers 17
    steel mast lines 18
    Trident construction 19
    types of overhead line support 16
    underground cables 22
    wood pole lines 19
nationalisation 9, 10
privatisation 11
REC 12
statutory framework 25
types of licence
    distribution 28
    generation 28
    inter-connector 28
    supply 28
    transmission 28

hearings
    application to Lands Tribunal
        appeals 205
        assessors 200
        consent orders 203
        expert evidence
            computer based valuations 199
            co-operation between 193
            discussions between experts 198
            duty of expert 192
            report 194
            use of opponent's evidence 196
            written questions to expert 196

managing the claim (the overriding objective)....................167
managing the claim (allocation of procedure) 167
negotiations, settlements and withdrawals 201
preliminary issues..189
pre-trial review 187
procedure
   background 157, 158
   reference to Lands Tribunal..................159, 160, 163, 164, 206
   rules 161, 163, 167
   working structure...158
sanctions..204
service of documents 165
site inspections..199
simplified procedure...182
special procedure 182
standard procedure
   award of costs
      detailed assessment 181
      principles 179
      procedures 178
      summary assessment 180, 181
   conclusion of reference 178
   decision 177
   definition...168
   documents..172, 175
   expert reports (direction for) 169
   final hearing 176
   further information about case 172
   interlocutory applications....................................170
   listing questionnaire 174
   reply 169
   statements of case...168
   witness statements (direction for) 169
   witness evidence 191
   written representations.......................................185
necessary wayleave hearings
   main hearing
      applications to Secretary of State......................30, 74, 79, 81
      conduct of 88, 90, 97
      evidence at (necessary and expedient).......................100
      evidence seved by licence-holders..........................94, 95
      representation at..96
      statutory rules governing 94
      submissions by objector 95

Index

  timing of 96
 post-hearing procedure
  costs 104
  decision of Secretary of State 103
  documenting grant of wayleave 104
  inspector's report 102
  no right of appeal 104
  re-opening of hearing 102
  time periods 103
 pre-hearing meeting
  adjournment 92
  agenda 93
  directions 93
  plans 94
  site inspection 99
  time periods 91
 public inquiry
  appointed inspector 122, 124
  consent review 128
  costs 129
  environmental issues 123
  format of inquiry 121, 126
  issues to be considered 122
  pre-inquiry meeting 123
 pre-inquiry procedure
  notice 122, 123
  qualifying objectors 124, 128
  timing 124
 procedure 122
 role of advocates 126
 role of inspector 124, 127, 128
 when held 122

land
 estates in
  legal 37, 38
  equitable 38
 interests in 38
 freehold 38
 leasehold 42
 rights 34, 43
 what is land 37, 66, 106, 148, 151
licences
 licence holders

```
      compliance with notice to remove ............................. 227
      definition................................................... 11, 28
      failure to comply with financial obligation ...................... 217
      identification of ........................................... 28, 60
      statutory substitution of ............................. 11, 12, 28, 57
      wrongful use of land by........................................ 231
  types of licence
      distribution................................................... 28
      generation..................................................... 28
      inter-connector................................................ 28
      supply......................................................... 28
      transmission................................................... 28

planning permission and ministerial consent
  overhead lines.................................................... 113
  Secretary of State consent
      application ................................................... 114
      consent ....................................................... 115
      objection ..................................................... 116
      review ........................................................ 118

remedies
  background....................................................... 209
  limitation period................................................. 215
  types
      damages ...................................................... 211
      debt ......................................................... 210
      injunction ................................................... 212
      mesne profits ................................................ 212

wayleaves
  ancillary rights
      associated environmental obligations .......................... 28, 71
      surveys for exploration ......................................... 49
      tree cutting ............................................. 48, 71, 82
      common legal features........................................... 39
  contrasted with other rights
      easements...................................................... 43
      tenancies...................................................... 42
  definition .................................................... 41, 44
  fees/payments
      calculation ................................................... 134
      contractual ..................................... 39, 137, 217, 218
      failure to comply financial obligations ......................... 217
```

348

```
    overhead lines through woodland............................. 138
    payments following necessary wayleaves............ 138, 157, 218, 221
    rent......................................................... 137
    underground cable rates ...................................... 136
    wayleave compensation code .............................. 140, 159
history of .................................................... 35, 36
nature of..................................................... 39, 44
need for
    adverse possession ............................................ 34
    easement................................................ 34, 43, 70
    Land Registration Act 2002 ..................................... 34
    prescription .................................................. 34
    right to enter land for maintenance ....................... 45, 47, 48
    trespass...................................................... 34
notices
    (see Application to Secretary of State for necessary wayleave)
    failure to comply with notice to remove ....................... 227
refusal to allow exercise of wayleave rights ....................... 223
registrability of ............................................... 42, 68
temporary continuation of .................................... 68, 227
termination of................................................... 75
types of......................................................... 52
    non-statutory voluntary wayleaves
        definition................................................ 52
        express................................................... 53
        voluntary wayleaves versus implied wayleaves........... 51, 52, 59
    non-statutory implied wayleaves
        arising of ................................................ 59
        identification of licence-holder ............................ 60
        terms.................................................. 59, 60
    necessary wayleaves
        definition........................................... 51, 66, 67
        (see Application to Secretary of State for necessary wayleave)
```

349

Printed in Great Britain
by Amazon